T0142476

Studies in Computational Intelligence

Volume 732

Series editor

Janusz Kacprzyk, Polish Academy of Sciences, Warsaw, Poland
e-mail: kacprzyk@ibspan.waw.pl

The series "Studies in Computational Intelligence" (SCI) publishes new developments and advances in the various areas of computational intelligence—quickly and with a high quality. The intent is to cover the theory, applications, and design methods of computational intelligence, as embedded in the fields of engineering, computer science, physics and life sciences, as well as the methodologies behind them. The series contains monographs, lecture notes and edited volumes in computational intelligence spanning the areas of neural networks, connectionist systems, genetic algorithms, evolutionary computation, artificial intelligence, cellular automata, self-organizing systems, soft computing, fuzzy systems, and hybrid intelligent systems. Of particular value to both the contributors and the readership are the short publication timeframe and the worldwide distribution, which enable both wide and rapid dissemination of research output.

More information about this series at http://www.springer.com/series/7092

Bruno Pinaud · Fabrice Guillet
Bruno Cremilleux · Cyril de Runz
Editors

Advances in Knowledge Discovery and Management

Volume 7

 Springer

Editors
Bruno Pinaud
University of Bordeaux
Bordeaux
France

Bruno Cremilleux
University of Caen Normandie
Caen Cedex 5
France

Fabrice Guillet
Polytech Nantes
Nantes
France

Cyril de Runz
University of Reims Champagne-Ardenne
Reims
France

ISSN 1860-949X   ISSN 1860-9503   (electronic)
Studies in Computational Intelligence
ISBN 978-3-319-88020-4   ISBN 978-3-319-65406-5   (eBook)
https://doi.org/10.1007/978-3-319-65406-5

Printed on acid-free paper

This Springer imprint is published by Springer Nature
The registered company is Springer International Publishing AG
The registered company address is: Gewerbestrasse 11, 6330 Cham, Switzerland

Preface

This volume features the extended versions of a selection of the best papers that were originally presented at the French-speaking conference EGC'2016 held in Reims (France) during 18–22 January, 2016. The papers have been accepted after a peer-review process among papers already accepted in long format at the conference. For the conference, the long papers are themselves the result of a double-blind peer-review process among the 94 papers initially submitted to the conference (conference acceptance rate for long papers of 24%). This conference was the 16th edition of this event, which takes place each year and which is now successful and well known in the French-speaking community. This community was structured in 2003 by the foundation of the International French-speaking EGC society (EGC in French stands for "Extraction et Gestion des Connaissances" and means "Knowledge Discovery and Management" or KDM). This society organizes every year its main conference (about 200 attendees) but also workshops and other events with the aim of promoting exchanges between researchers and companies concerned with KDM and its applications in business, administration, industry, or public organizations. For more details about the EGC society, please consult http://www.egc.asso.fr.

Structure of the Book

This book is a collection of six representative and novel works done in Data Mining and Knowledge Discovery. It is intended to be read by all researchers interested in these fields, including Ph.D. or M.Sc. students, and researchers from public or private laboratories. It concerns both theoretical and practical aspects of KDM.

Alec et al. propose in Chapter "A Combined Approach for Ontology Enrichment from Textual and Open Data" an approach for ontology enrichment for automatically labeling documents describing entities, with very specific concepts reflecting specific users' needs. This new approach to ontology population and enrichment exploits the foundations of the Semantic Web by combining contributions of text analysis, linked open data extraction, machine learning, and reasoning tools. An evaluation, on two application domains, provides quality results and demonstrates the interest of the approach.

Chapter "C-SPARQL Extension for Sampling RDF Graphs Streams", by Fall Dia et al., presents an extension of the semantic data stream processing system C-SPARQL for sampling RDF graphs streams. The authors add new operators to C-SPARQL query syntax and propose a solution based on existing sampling algorithms (Uniform, Reservoir, and Chain) for reducing load while keeping data semantics. The evaluation shows an improved response time of query execution and a perfect preserving of data semantics.

Chapter "Efficiency Analysis of ASP Encodings for Sequential Pattern Mining Tasks", written by Guyet et al., presents how the declarative paradigm of Answer Set Programming may be used to define the foundations for developing a flexible sequential pattern mining tool. It proposes encodings of the classical sequential pattern mining tasks within two representations of embeddings and for various kinds of patterns. The computational performance of these encodings is compared each other to get a good insight into the efficiency of ASP encodings and to a constraint programming approach.

In multi-writer, multi-reader systems, data consistency is ensured by the number of replica nodes contacted during read and write operations. Chapter "Consistency-Latency Trade-Off of the LibRe Protocol: A Detailed Study", by Kumar et al., develops a consistency protocol called LibRe, which helps to read the latest version of a data item by contacting a minimum number of replica nodes. This protocol provides a new trade-off between consistency and latency for distributed data storage systems. The effectiveness of the approach in practice is shown by providing a proof-of-concept implementation of the protocol inside the Cassandra distributed data store.

Taleb et al. in Chapter "A Fault Detection Approach for Robotic Systems Combining the Data Obtained from Sensor Measurements and Linear Observer-Based Estimations" propose a new approach for fault detection in robotic systems which is performed by using classification algorithms. In the first stage, data are collected from the different sensors installed on a real system. A linear state observer is used to estimate non-measurable state variables in order to enrich the database. Then, a classifier is used to diagnose the faults occurring on the considered system. An experiment on a robotized actuated seat is considered to illustrate the proposed fault detection approach and demonstrate the benefits of adding estimations of unmeasured state variables to the classification process.

Chapter "A Collaborative Framework for Joint Segmentation and Classification of Remote Sensing Images" by Troya-Galvis et al. presents a collaborative framework for joint segmentation and classification which is quality-centric and attempts to simultaneously improve both segmentation and classification for a given thematic class by the interaction of these two paradigms. The framework is applied to vegetation extraction in a very high spatial resolution image of Strasbourg, France. The experiments show that the proposed method reaches good classification results while remarkably improving the segmentation results.

Acknowledgements

The editors would like to express their gratitude to the chapter authors for their insights and contributions to this book.

The editors would also like to acknowledge the members of the review committee and the associated referees for their involvement in the review process of the book. Their in-depth reviewing, criticisms, and constructive remarks have significantly contributed to the high quality of the selected papers.

Finally, we thank Springer and the publishing team, and especially T. Ditzinger and J. Kacprzyk, for their confidence in our project.

Bordeaux, France   Bruno Pinaud
Nantes, France   Fabrice Guillet
Caen, France   Bruno Cremilleux
Reims, France   Cyril de Runz
June 2017

Organizing Committee

Review Committee

All published chapters have been reviewed by 2 or 3 referees and at least one not native French speaker referee.

- Mohammed Bouguessa (Université du Québec à Montréal (UQAM), Canada)
- Toon Calders (Université libre de Bruxelles, Belgium)
- Francisco de A.T. De Carvalho (Universidade Federal de Pernambuco, Brazil)
- Carlos Ferreira (LIAAD INESC Porto LA, Portugal)
- Antonio Irpino (Second University of Naples, Italy)
- Daniel Lemire (LICEF Research Center, University of Québec, Canada)
- Paulo Maio (GECAD—Knowledge Engineering and Decision Support Research Group, Portugal)
- Fionn Murtagh (University of Derby; Goldsmiths University of London, UK)
- Benoit Otjacques (Luxembourg Institute of Science and Technology)
- Luiz Augusto Pizzato (1-Page, Australia)
- Lorenza Saitta (Università degli Studi del Piemonte Orientale, Italy)
- Dan Simovici (University of Massachusetts Boston, USA)
- David Taniar (Monash University, Australia)
- Stefan Trausan-Matu (University Politehnica of Bucharest, Romania)
- Jef Wijsen (University of Mons-Hainaut, Belgium)

Associated Reviewers

Herman Akdag, Yousra Chabchoub, Pierre Gancarski, Tom Lampert, René Quiniou, Prajwol Sangat Olivier Hermant

Contents

About the Editors

Bruno Pinaud received the Ph.D. degree in Computer Science in 2006 from the University of Nantes. He is currently Assistant Professor at the University of Bordeaux in the Computer Science Department since September 2008. His current research interests are visual data mining, graph rewriting systems, graph visualization, and experimental evaluation in HCI (human–computer interaction).

Fabrice Guillet is a Full Professor in CS at Polytech Nantes, the graduate engineering school of University of Nantes, France, and a member of the "Data User Knowledge" team (DUKe) of the LS2N laboratory. He received a Ph.D. degree in CS in 1995 from the "École Nationale Supérieure des Télécommunications de Bretagne", and his Habilitation (HdR) in 2006 from Nantes University. He is a co-founder and president of the International French-speaking "Extraction et Gestion des Connaissances (EGC)" society. His research interests include knowledge quality and knowledge visualization in the frameworks of Data Science and Knowledge Management. He has co-edited two refereed books of chapter entitled "Quality Measures in Data Mining" and "Statistical Implicative Analysis—Theory and Applications" published by Springer in 2007 and 2008.

Bruno Cremilleux is professor in computer science since 2005 at the University of Caen-Normandy, France. He received his Ph.D. in computer science in 1991 from the University of Grenoble. His main research interests are pattern (set) discovery, Constraint Satisfaction Problems and data mining, preference queries, and exploratory data mining. This research work benefits from close collaborations addressing applications mainly in the fields of Chemoinformatics and Biomedical Text Analysis.

Cyril de Runz is a lecturer at the University of Reims Champagne-Ardenne, France. He obtained his Ph.D. in computer science in 2008 and his Habilitation in 2015 from the same university. His fields of interests are artificial intelligence, data mining, fuzzy set theory, geomatics, and information systems.

A Combined Approach for Ontology Enrichment from Textual and Open Data

Céline Alec, Chantal Reynaud-Delaître and Brigitte Safara

Abstract This paper proposes an approach for ontology enrichment for automatically labeling documents describing entities, with very specific concepts reflecting specific users' needs. The peculiarity of this approach is that it addresses a triple challenge: (1) the concepts used for labeling have no direct terminology in the documents, (2) their formal definitions are not initially known, (3) the information useful to label the documents is not necessarily mentioned in them. To solve those problems, we propose to use an existing ontology of the domain of concern and to enrich it with the definitions of the concepts used for labeling. To construct these definitions, we work on a set of manually labeled documents, used as examples. The ontology is populated with information extracted from these documents, and with information coming from external resources (Linked Open Data). The definitions that we want to get can then be learned based on this populated ontology and on the set of labeled documents. Learned definitions are then added to the ontology (ontology enrichment). Hence, whenever new documents of the same domain have to be labeled, the ontology can be populated in the same way and definitions apply, allowing the new documents to be labeled. This approach, named SAUPODOC, is a novel approach to ontology population and enrichment, exploiting the foundations of the Semantic Web by combining contributions of text analysis, linked open data extraction, machine learning and reasoning tools. An evaluation, on two application domains, provides quality results and demonstrates the interest of the approach.

C. Alec (✉) · C. Reynaud-Delaître · B. Safara
LRI, Univ. Paris-Sud, CNRS, Université Paris-Saclay, 91405 Orsay, France
e-mail: celine.alec@lri.fr

C. Reynaud-Delaître
e-mail: chantal.reynaud@lri.fr

B. Safara
e-mail: brigitte.safar@lri.fr

© Springer International Publishing AG 2018
B. Pinaud et al. (eds.), *Advances in Knowledge Discovery and Management*,
Studies in Computational Intelligence 732, https://doi.org/10.1007/978-3-319-65406-5_1

1 Introduction

This work is the result of a collaboration between LRI (Laboratoire de Recherche en Informatique) and the Wepingo[1] company, a startup which develops on-line applications for proposing products to web users. Our goal is to design an approach for automatically labeling documents describing products (or more generally entities), with very specific concepts corresponding to specific user needs. This aims at facilitating the design of flexible systems that can be adapted to different product categories and different points of view on these products. The specificity of our approach is that it is faced with three problems: (1) the documents do not contain word for word the label of specific concepts; (2) each specific concept is expressed as a label, i.e., we do not have any formal definition, but the system designer knows what type of data must be found in order to label the documents with the specific concepts; (3) documents are incomplete, they may not contain all the information necessary for this labeling.

Our problem can be exemplified by the tourism domain. Let us suppose that we have a corpus of documents, each document describing a touristic destination. We would like to find out, for each destination (so, for each document), whether the destination is a *DWW*, i.e., "Destination where you can practice Watersports during Winter", or not. This concept is very specific and never appears as such in the documents describing the destinations. However, the system designer knows that the concept refers to a place that is warm enough in winter and with watersports. He/she also knows that one can find in the documents terms referring to watersports but he/she will not find the temperature ranges in winter, and he/she will have to look for them in an external resource.

Once the necessary information is acquired, a description can then be labeled as an instance of the concept *DWW* or not. Automating the process requires clarifying the formal definition in an appropriate language, which can be difficult to achieve: for instance, what does a place warm enough in winter explicitly mean? However, a formal definition can be learned by an automatic tool if the system designer provides a number of descriptions that are already labeled as positive or negative examples of the concept. Once the definition has been learned, new descriptions of destinations satisfying this definition can be automatically labeled as instances of the concept (positive labels) or as not instances of the concept (negative labels).

Explicitly knowing the definitions gives more flexibility to the system. If no destination complies with a given definition, some constraints on the definition can be deleted in order to propose at least a few destinations. For example, a user seeking *DWW* will eventually be satisfied with a destination where the temperatures are slightly lower than what is expected.

The contribution of this paper is an original approach for ontology enrichment that combines different text analysis tools, LOD (Linked Open Data) extraction, machine learning and reasoning, in a context constrained by the three statements outlined above. The rest of the article is organized as follows. Section 2 presents the

---

[1] http://www.wepingo.com/.

state of the art. Section 3 describes the approach and Sect. 4 details its various tasks. Section 5 presents the experiments. We conclude in Sect. 6 on future work directions.

2 Related Work

Ontology enrichment is a vast field of research in which we distinguish three categories of works that focus on the extraction of semantic knowledge from more or less structured texts.

The first category concerns works on expressive ontologies and generation of concept definitions. Some approaches work on texts describing concepts. For example, Lexo (Völker et al. 2007) applies syntactic transformation rules on natural language definitions for generating axioms in Description Logic (DL). Ma and Distel (2013b) uses an approach based on the extraction of relations and relies on formal constraints to ensure the quality of the learned definitions (Ma and Distel 2013a). These approaches are not applicable on the documents that we deal with because they contain only descriptions of instances. Others, such as Chitsaz (2013) and Lehmann and Hitzler (2010), have only descriptions of instances, just like us. They rely on inductive logic programming to find new logical descriptions of concepts from the assertions of an ontology. Lehmann and Hitzler (2010) applies to expressive ontologies in DL, whereas Chitsaz (2013) applies to lightweight ontologies and both require a large number of assertions related to the instances. In comparison, our inputs are incomplete and weakly structured texts, from which assertions have to be extracted.

The second category of works deals with the generation of lightweight ontologies like taxonomies. They study how to extract different ontological elements from textual resources (Cimiano 2006). For the extraction of concepts, the key step is the extraction of the relevant terminology of the domain (Cimiano et al. 2006) using different measures of term weighting. Classification techniques are then applied to detect synonyms and an ontological class can be derived for each group of similar terms. Others are interested in learning concept hierarchies. They mainly use unsupervised hierarchical classification techniques to simultaneously learn the concepts and their subsumption relations (Cimiano 2006). Finally, supervised methods can be used in the case where an existing concept hierarchy has to be extended with new concepts. Classifiers must be trained for each of the existing ontology concepts which must not be very large. Appropriate similarity measures are then used to compare a new concept with the existing ones (Cimiano and Völker 2005). All these works are designed to recognize words denoting concepts (or instances) in the texts and then extract them. However, sometimes the texts only mention the properties of the instances without naming the underlying concept, as in our work. Other approaches, such that those described below, are then necessary.

The third category includes works that use reasoning to partially replace traditional extraction techniques. In the BOEMIE system (Petasis et al. 2013), concepts are divided into primitive and composite concepts, the latter ones being defined from the first ones. The primitive concepts are populated with standard extraction tools.

Instances of the composite concepts are not explicitly present in the texts but their properties are. The composite concepts are populated by reasoning on the extracted properties and on instances of primitive concepts. Yelagina and Panteleyev (2014) extracts facts from texts thanks to natural language processing tools and an ontology. From these facts, background knowledge and inference rules introduced beforehand, new facts not mentioned in the text can be derived. Our work is close to these two works but differs in the fact that we do not have definitions of the concepts to be populated.

This state-of-the-art shows that none of the approaches taken in isolation is a solution to our problem. The next section describes the approach SAUPODOC that combines some processes to fit our context.

3   The Principle of the SAUPODOC Approach

3.1   Description of the Approach

Our approach, called SAUPODOC (Semantic Annotation Using Population of Ontology and Definition Of Classes) is automatic and generic. This means that the same approach can be used for several domains, provided that all the inputs of the framework are adapted to the concerned domain. The approach aims at labeling input documents with so called target concepts. For each target concept, a document must be labeled either as an instance of this concept or as not an instance of this concept.

The SAUPODOC approach is composed of two workflows. The first one, performed once, consists in learning definitions from example documents, manually annotated. The second one consists in applying the learned definitions to new documents that need to be labeled. These two workflows rely on four tasks which are guided by a domain ontology. The first two tasks, used by both workflows, consist in populating the ontology by extracting the data associated with the described entities. This extraction is performed from the textual documents (Task 1) and external resources (Task 2). The other tasks are reasoning tasks on the populated ontology: the discovery of formal definitions of target concepts in the first workflow (Task 3) and the population of these target classes on the second workflow leading to the labeling of the documents (Task 4).

The structure of both workflows is illustrated by Figs. 1 and 2. During the whole process the ontology, initially denoted by O is progressively enriched.

The first workflow aims at constructing the definitions of target concepts. The first task populates the ontology with property assertions. These are extracted from the information contained in documents of an input corpus. Then, the second task extends these assertions using information extracted from an external resource. The populated ontology (O^+) is then enriched. The documents of the input corpus are manually labeled, in order to obtain positive and negative examples for each target concept.

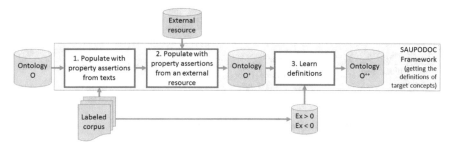

Fig. 1 Workflow 1 of SAUPODOC: getting the definitions of target concepts

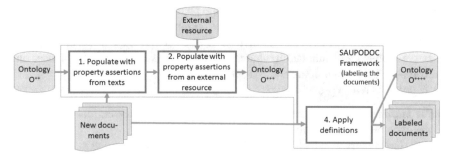

Fig. 2 Workflow 2 of SAUPODOC: labeling the documents

Definition of those target concepts are learned automatically from the examples and then inserted as classes into the ontology (O^{++}).

The O^{++} ontology can then be used for labeling new documents, cf. Figure 2. To do that, Tasks 1 and 2 have to be done, in the same way as for the labeled corpus, in order to populate the ontology with property assertions describing the entities of the new documents (O^{+++}). Finally, the definitions of the target concepts are applied, populating the classes corresponding to the target concepts (O^{++++}) and labeling the new documents with the target concepts.

The next subsections describe the inputs of the approach: (1) a domain ontology, assumed to be previously defined, enriched by the properties that the system designer believes to be involved in the future definitions of target concepts, and (2) a corpus describing entities in the form of textual descriptions, labeled as positive or negative instances of each target concept.

3.2 The Input Ontology

The input ontology defines the domain under consideration. It is a guide to the analysis of documents, the search for additional information and the reasoning on the obtained definitions and assertions. This means that the ontology has to contain

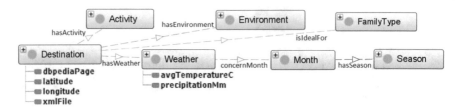

Fig. 3 Structure of the ontology about touristic destinations

all the elements defining the domain entities. It is domain-specific but not approach-specific. Its constitution is not the focus of the paper.

More formally, the ontology O is an OWL (Ontology Web Language) ontology defined as a tuple (C, P, I, A, F) where C is a set of classes, P a set of properties (datatype, object and annotation) characterizing the classes, I a set of instances and their types, A a set of axioms and F a set of facts.

C contains two types of classes:

- **the main class**, which corresponds to the general type of the entities described in the documents. For example, Destination in Fig. 3.
- **descriptive classes**, which are all other classes, used to define the main class. For example, Activity or Environment in the Fig. 3.

Figure 3 presents an excerpt of an ontology about touristic destinations where the main class is Destination. Descriptive classes Activity, Environment, FamilyType and Season are respectively hierarchy roots. For example, Environment represents the natural environment (Aquatic, Desert, etc.) or its qualities (Beauty, View). Some of the inter-class properties have sub-properties, which are not shown here. Datatype properties appear under the classes.

A contains constraints on the set of classes C or on the set of properties P such as subsumption, equivalence, disjunction, domain/range as well as characteristics (functional, transitive, etc.).

At first I contains only instances of descriptive classes. For instance, _rainForest is an instance of Forest (descendant of Environment) and dense forest is one of its labels.

The set of facts F is initially empty. It will contain property assertions, i.e., triples like <Dominican_Republic hasEnvironment _rainForest> that will be introduced by Tasks 1 and 2.

The set of target concepts used by the system designer to label documents will be introduced by Task 3 as a set of specializations of the main class. This set will not form a partition, as a product may be an instance of several target concepts.

3.3 The Labeled Corpus

The labeled corpus consists of a set of XML (Extensible Markup Language) unstructured documents. Every document describes a particular instance of the main class, i.e., a product, or more generally an entity. The structure reveals both the name of the instance and the textual description. The textual description contains labels of instances of descriptive classes from the ontology but no label of the target concepts, which can be complex sentences like "Destination where you can practice Watersports during Winter" (*DWW*). Each document is manually labeled by the system designer, as a positive or negative instance of each target concept. Note that in our context, the documents are either extracted from advertising catalogs (and they praise the virtues of the entities described) or very short descriptions. In all cases, few negative expressions are present and they mention all the characteristics of entities. However, some specific information may be missing like numerical values.

For the second workflow, the new documents have the same features as those of the labeled corpus (XML documents describing all the main characteristics of entities and with few negative expressions). But, they are not labeled.

4 Tasks Implemented in the SAUPODOC Approach

This section details each task involved in the approach. Note that, because it is successively enriched by the different tasks, the ontology here plays a central role, both as unifying language supporting some kind of task cooperation, as well as a guide for each of them, since these are driven by its terminology and structure.

4.1 Preliminary Task

In a preliminary task, entities described by each document of the corpus are added to the ontology as individuals of the main class. For example, if the main class of the tourist destinations ontology is Destination and if the corpus contains a document "Dominican_Republic.xml", then the ontology is populated with an individual Dominican_Republic such as <Dominican_Republic, rdf:type, Destination>.

4.2 Task 1: Population with Property Assertions from Texts

In the first task, data is extracted from textual descriptions of entities by a document annotation process according to the domain ontology. We chose to use the GATE annotation software (Bontcheva et al. 2004; Cunningham et al. 2011) because it

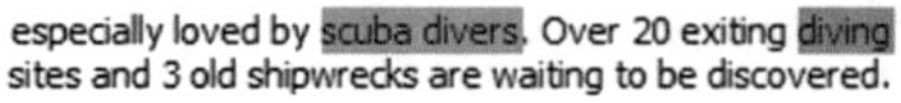

especially loved by scuba divers. Over 20 exiting diving
sites and 3 old shipwrecks are waiting to be discovered.

Fig. 4 Extract from the document about the Dominican Republic annotated by GATE

performs different text analysis tasks allowing the user to choose the ontology which
will guide the process. Other tools like Open Calais[2] are not able to do that. Figure 4
is a snippet of the document describing the Dominican Republic destination, the
terms scuba divers and diving are associated by GATE with the individual _diving from
the ontology, instance of a concept specializing WaterSport.

Property assertions for each entity that are instances of the main class are then
populated thanks to GATE annotations. The JAPE language (Java Annotation Patterns
Engine), usable with GATE, is able to define rules transforming annotations into
property assertions in the ontology. We have built a generic pattern, suitable for
any ontology, which is automatically converted into as many JAPE rules as ontology
properties to be populated. The system designer has to specify in the ontology what
properties should be populated by the JAPE rules. For example, he or she can state
that hasActivity must be considered but not hasWeather. For each property, the process
is guided by the constraints on ranges expressed in the ontology. For example, the
constraint <Destination, hasActivity, Activity> requires the range value taken by the
property hasActivity to belong to the extension of the class Activity. From this constraint,
if the description of an entity e contains an annotation corresponding to an instance
a of Activity, then the assertion <e, hasActivity, a> is built. In the example of Fig. 4, as
_diving is an instance of a concept specializing WaterSport (a subclass of Activity) the
assertion <Dominican_Republic, hasActivity, _diving> is added.

In our context, where documents praise the advantages of the described entities
and therefore contain no (or few) negative expressions, this simple population process
appears to be sufficient.

4.3 Task 2: Population with Property Assertions from an External Resource

Task 2 aims at completing the property assertions found in Task 1. Indeed, the documents are often incomplete. They usually do not contain all the information needed
to define a target concept. For example, intuitively the definition of a *DWW* takes
into account the weather of destinations in winter. The temperature and precipitation
per season or per month do not appear in the descriptions. The collection must be
enriched by exploiting resources available on the Web.

To perform this task, the system designer needs an RDF resource containing
entities of the domain of the corpus and to identify in this resource the properties
that are required by the ontology. Currently, the resource used is DBpedia.

---

[2]http://www.opencalais.com/.

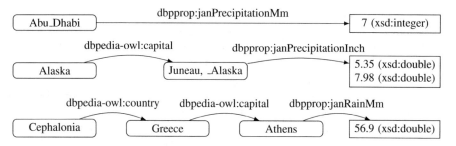

Fig. 5 Exploration paths in DBpedia

The task has two steps. First, we use DBpedia Spotlight (Mendes et al. 2011). This tool can automatically annotate the references to DBpedia entities in a text. We apply it to the name of the instance of each document of the corpus. It allows us to have a direct access to the DBpedia page representing the entity of the document. Other tools could have been used like Wikifier (Cheng and Roth 2013; Ratinov et al. 2011) or AIDA (Yosef et al. 2011), but the access would not have been direct since they return Wikipedia pages instead of DBpedia pages. The second step consists in automatically generating SPARQL queries to add property assertions to the ontology. These queries are applied via the DBpedia SPARQL endpoint.[3] A summary of this task, detailed in Alec et al. (2016) is given below.

We face a correspondence problem. Indeed, the vocabularies of the ontology and of the external resource like DBpedia may differ. Mechanisms must be conceived to establish mappings between the required ontology properties and those of the resource. Various mechanisms have been set up to deal with cases where, for example, a source property (from the ontology) corresponds to several equivalent target properties with different syntaxes. For instance, the property precipitation_in_January is represented by six different properties in DBpedia (janPrecipitationMm, janRainMm, janPrecipitationInch, janRainInch, janPrecipitationIn and janRainIn). Other mechanisms not listed here allow us to match with target properties that are calculated or obtained following an aggregation process.

We also take into account the incompleteness of DBpedia and therefore the fact that there could be no known value for some property. Indeed, in our context of definition learning from examples, the data extracted from examples should be as complete as possible, because it impacts the quality of the learned definitions. We have therefore developed a mechanism for exploring nearby pages in the resource graph in order to obtain a value that can serve as an acceptable approximation of a missing property. Figure 5 shows two examples (Alaska and Cephalonia) where the value of the property for the January precipitation is not on the destination page itself but where an approximate value can be found on a nearby page (the page of the capital city).

---

[3]http://dbpedia.org/sparql.

This mechanism is based on the composition of properties and allows the designer to establish alternative access paths of the DBpedia graph, until the pages containing the required information are reached. SPARQL queries (with the CONSTRUCT query form) based on these specifications collect the data and create property assertions that will be inserted into the ontology. Note that, for object properties, property assertions are not the only thing to be added. The ranges of these assertions (individuals) are added to the ontology as instances of the range of the object property under consideration. For example, let us suppose that we have an ontology about films with an object property hasLanguage such as its range is the class Language. Suppose Task 2 adds the assertion < A_Crime_in_Paradise, hasLanguage, _French_language > regarding a film named "A_Crime_in_Paradise" described in a document. In this case, an individual _French_language is created such as < _French_language, rdf:type, Language >.

4.4 Task 3: Definition Learning

Task 3 is a learning task that is performed only once, in the first workflow. First, the target concepts are inserted in the ontology as subclasses of the main class, that we call target classes. Then, definitions of each target concept are learned using the tool DL-Learner (Lehmann 2009). To learn definitions, this tool uses the ontology and positive and negative examples of the different target concepts. These definitions are added to the ontology via axioms of equivalence between a target class and its definition.

DL-Learner has been chosen for this task because it is an open source tool using an input ontology. It is able to learn class definitions expressed in Description Logic. To the best of our knowledge, two other existing tools perform the same task, YinYang (Esposito et al. 2004) and DL-FOIL (Fanizzi et al. 2008), but they are not open source. DL-Learner has many advantages in our context compared to most machine learning tools. First, it allows us to obtain the explicit definitions of each target class, which is an important asset in concrete applications. Second, most machine learning tools do not take into account the relations explained in an ontology (subsumption, object properties across classes) in their representations of the examples.

The definitions built by DL-Learner are conjunctions or disjunctions of items. An item can be a class (Destination), an expression about an object property at the class level (hasActivity some Nightlife), an expression about an object property at the instance level (hasActivity value _diving), an expression about a datatype property (avgTemperatureC some double[>= 23.0]), or a cardinality constraint (hasCulture min 3 Culture). The ranges can also be conjunctions or disjunctions of items. Hence, with a sufficient number of labeled examples given as inputs, a definition can be learned. For example, the definition of the class *DWW* can be something like this:

(Destination and (hasActivity some Watersport)
    and (hasWeather min 2 ((concernMonth some (hasSeason some MidWinter))
        and (avgTemperatureC some double[>= 23.0])
        and (precipitationMm some double[<= 70.0]))))).

To setup DL-Learner, we have used the CELOE (Lehmann et al. 2011) (Class Expression Learning for Ontology Engineering) algorithm announced as the best for learning classes, and the default reasoner that uses the closed-world assumption (CWA). However, we have disabled the negation (NOT) and universal restriction (ONLY) operators since learned definitions should be introduced in the OWL ontology and reasoning in it is under the open-world assumption (OWA). In addition, in order to learn and operate with minimum cardinality constraints such as (hasActivity min 3 Activity), instances have automatically been expressed as disjoint (Unique Name Assumption) not to be assumed likable via an owl:sameAs. However, as the reasoning cannot work under OWA with definitions containing maximum cardinality restrictions (MAX, EXACTLY), they have been ignored, i.e., we automatically retain the best definition that does not contain such restrictions. In addition to the basic configuration, corresponding to the parameters described above, we have defined a complex configuration. A search heuristic is activated so that we can learn definitions that are really complex and far from easy to learn, such as the one expected for *DWW*.

A final important parameter of DL-Learner is the noise (called noisePercentage in DL-Learner), i.e., the percentage of positive examples that we accept to be not covered by the definitions. We proceeded by trial and error to adjust it and have established a methodology based on the conducted experiments. For every target class, 10 configurations are tested: the basic and complex configurations, each with 5 different noise values (5-15-25-35-45%). For each configuration, the highest ranked solution in terms of accuracy and size was chosen, then for each target class, the definition chosen is the best out of the 10.

At the end of this step, we have an ontology enriched by the definitions of target concepts. These definitions will be used to label new documents of the same domain.

4.5 Task 4: Definition Application

The output of Task 3 is an ontology including the definitions of target concepts. As described in Fig. 2, when we have some new documents of the same domain, we perform Tasks 1 and 2 on these documents in order to have property assertions for the entities. Then, Task 4 can be performed. It consists in applying the definitions to populate the target classes in the ontology. For this task, we chose to use FaCT++ (Tsarkov and Horrocks 2006), an available OWL-DL reasoner. We also tried to use HermiT (Shearer et al. 2008) and Pellet (Sirin et al. 2007) but they failed when tested on an ontology with lots of individuals (more than 10,000 individuals). FaCT++ applies the definitions of target classes thanks to the property assertions known about the entities. Hence, it is able to identify the entities that correspond to a given definition. This task populates the target classes with the individuals complying with their definitions. For a given target class tc, if the entity described in a document d complies with the learned definition for tc, then it is recognized as an instance of tc. Hence, the document d is labeled by tc, otherwise it is labeled by *not tc*.

By doing this, we simulate a CWA while OWL reasons under OWA. Our particular context allows us to simulate the CWA at all steps. Hence, as we indicated in Sect. 4.4, definitions are built under CWA, from property assertions extracted from the examples. For an entity, when a property cannot be extracted, we assume it does not hold. For example, if a description of destination does not mention any beaches, we are sure that there are no beaches, since the documents are supposed to mention all the characteristics of the places. Similarly, to deal with the incompleteness of DBpedia data, we defined a path model to replace missing values by approximations in order to have all the required data.

5 Experiments

We have compared the approach SAUPODOC with two classification approaches, one based on SVM (Support Vector Machine) and the other on a decision tree. The experiments have been made on two application domains described below.

5.1 Materials

5.1.1 The Domain of Touristic Destinations

The corpus of touristic destinations contains only 80 documents. It is quite small, which makes a manual check of the found property assertions possible. Each document is automatically extracted from the catalog of the Thomas Cook website[4] and describes a particular destination (country, region, city or island). Documents are promotional, i.e., they highlight the qualities of destinations and contain very few negative expressions. Geolocation and weather data are missing. This information will be extracted from DBpedia through Task 2.

The domain ontology is well-structured. It includes a main class Destination and 161 descriptive classes. These are used to characterize the nature of the environment (46 classes), the possible activities (102 classes), the type of concerned families, e.g., with children, couples, etc. (6 classes) and classes for the weather (7 classes). Descriptive classes contain instances and their terminology forms for an easy identification in the texts. For example, the terms archaeology, archaeological, acropolis, roman villa, excavation site, mosaic are associated with the instance archaeology. 39 target concepts are under consideration.

---

[4]http://www.thomascook.com/.

5.1.2   The Domain of Films

The corpus of films contains 10,000 documents. It is quite large, which allows us to verify the applicability of the approach with many individuals. This corpus has been automatically built from DBpedia. Each document corresponds to a DBpedia page of a film, and contains the film name, the URI of this page (the page is already known, DBpedia Spotlight will not be used for the film study case) and a summary of the film (with very few negative expressions). The duration of the film, its languages and countries of origin will be extracted from DBpedia. Indeed, the duration of the film is not mentioned in the descriptions while the languages and countries of origin may be mentioned, but a misinterpretation is possible. For example, the word "French" in the summary may have various meanings: the film may be French (country) or in French (language), or it can tell the story of a French. We therefore prefer to use the information from DBpedia, which is clearly stated as language or country of origin, rather than the information from texts.

The ontology dealing with films is very simple, with little structure. It contains the main class Film and only the 5 descriptive classes useful for the target concepts of our experiments. If new target concepts appear, the ontology would have to be adapted w.r.t. them. Both descriptive classes Language and Country initially have no instances, since Task 2 is able not only to add assertions but in addition to instantiate descriptive classes. The 12 chosen target concepts correspond to categories of DBpedia, given by the property dcterms:subject, which allows us to automatically obtain the labeled examples of each target concept.

5.2   Experimental Scenario

The positive and negative examples of each target concept must be given as input for each tested approach. They are given by the designer of the application in the case of destinations and automatically generated for films: a film f is a positive example for a target concept corresponding to the class c when it has the property $< f$ dcterms:subject $c >$, and a negative example otherwise.

SAUPODOC is based on an ontology, in contrast to the classifiers we are comparing with (SVM and Decision tree). We use the terminology of the ontology as a domain dictionary. What we call the terminology of the ontology is the set of individuals that are instances of descriptive classes, including all of their labels. Thus, one word from this dictionary corresponds to one individual from the ontology. All of its labels are taken into consideration as if they were a unique same word.

Each document is modeled by a vector (Vector Space Model), with a bag-of-words representation where each element of the vector corresponds to a word in the dictionary (that can be one or several keywords). After a lemmatisation phase, when a document contains a word, the value of the corresponding element of the vector is the TF-IDF (Term Frequency-Inverse Document Frequency) value, otherwise the value is 0. The resulting vector representation is used as inputs of the two classifiers.

The classifiers are tested with several parameters and the best results are kept. For the evaluation, the set of annotated documents is split into two sets: the training set which contains two thirds of documents and the test set containing the remaining documents. This means that the learning is done on the two thirds of documents and the results are assessed on the remaining documents. Several metrics are calculated.

5.3   Results on the Test Set

Table 1 shows the accuracy, F-measure, precision and recall of our approach and of the two classifiers. Let us focus first on the accuracy. The three approaches work well, even if SAUPODOC is a bit better. Accuracy is important to see the overall correct labeling but it is not the only measure to be considered in our context because each target class has mainly negative examples. For example, over the film target classes, the average percentage of negative examples is 91.76%. This means that a simple classifier predicting only negatively for all inputs has a high accuracy. Other measures such as precision, recall and F-measure are thus needed to make an assessment of the positive prediction, which is important in our context. We can see in Table 1 that our approach is better than the two classifiers on these measures.

$$Accuracy = \frac{TP + TN}{TP + FP + TN + FN} \qquad Precision = \frac{TP}{TP + FP}$$

$$F - measure = \frac{2 \times precision \times recall}{precision + recall} \qquad Recall = \frac{TP}{TP + FN}$$

Our results combine the performance of the different tasks performed by SAUPODOC. The learning task (Task 3) allows a good classification, but the previous tasks (Tasks 1 and 2) impact the results as well. Indeed, they play a role in the quality of the data used to learn the definitions. Since the touristic destination corpus is small, we performed a manual assessment of these two first tasks.

In Task 1, property assertions are created thanks to the information in the textual descriptions. 2,375 property assertions are created. Among them, 52 are false (false positives). This means the precision of this task is 97.81%. We do not calculate

Table 1 Average results for destinations (39 target classes) and films (12 target classes). "Us" describes the results for SAUPODOC whereas "SVM" and "Tree" respectively describe the results for the SVM classifier and the Decision tree classifier

Metric	Accuracy (%)			F-measure (%)			Precision (%)			Recall (%)		
Corpus	Us	SVM	Tree	Us	SVM	Tree	Us	SVM	Tree	Us	SVM	Tree
Dest.	95.89	84.52	86.23	72.23	54.14	63.22	73.95	58.10	64.23	71.58	55.32	65.89
Film	95.46	94.41	94.32	75.65	61.74	61.40	76.27	69.90	67.72	77.76	57.59	58.99

the recall, because if a property assertion is not mentioned in the text, then this property does not characterize the instance described since all the important features are assumed to be included in the descriptions. This means our context supposes the recall to be equal to 1, since there are no false negatives (missing assertions). Task 1 does not bring much nose to the ontology.

For Task 2, the techniques proposed to address the multiple or multi-valued or missing properties have proved their utility. In our tests, performed on DBpedia 2014, only 29 of the 80 touristic destinations have the desired weather data. The specification of access paths has allowed approximate values to be found: for example, temperatures for Boston have been obtained from the page Quincy,_Massachusetts.

Moreover, our approach has another benefit over the classifiers. These ones do not provide any explicit definitions. SVM classifiers generate a model that is not understandable by a human being. Decision trees are a little more understandable since they are sets of rules, but here these rules refer to TF-IDF values, so their human interpretation is not obvious. In SAUPODOC, definitions are directly understandable by both a human and a tool. This means a refinement work can be performed as a post-process (see future works in Sect. 6).

5.4 Validation of the Approach

The labeled corpus has been divided into two thirds of documents for the training set and one third for the test set. In our approach, as well as for classifiers, a number of parameters are tested and we only keep the best results, that is, those that generate the best accuracy on the test set. Thus, a slight bias is introduced into the results as they represent the models generating the highest accuracy on the test set, and then potentially induces a drop on all measures with a new set (accuracy, precision, recall, F-measure).

Since we do the same in the three approaches (SAUPODOC, SVM, decision tree), we consider that the differences between the obtained results are sufficiently characteristic. To verify this, we have used a new labeled sample in the domain of films. This sample, called validation set, contains 10,000 new documents labeled with the 12 film target concepts. It has been obtained in the same manner as our training and test sets, i.e., using DBpedia. We only do this experiment with the film domain because a validation set is easy to create on this domain thanks to DBpedia.

On this validation set of 10,000 documents, we apply our approach with the definitions obtained in the experiments of Sect. 5.3 and we apply the classifiers obtained too. In Fig. 6, we can observe that the same trends emerge for the test set and the validation set. Therefore, the experiments of Sect. 5.3 make sense.

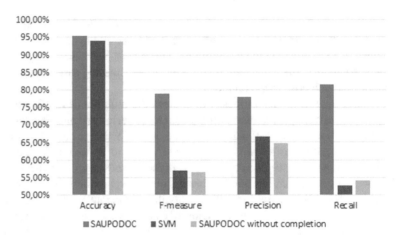

Fig. 6 Results on the validation set (10,000 films)

5.5   *Experiments and Discussion on Completion*

A big advantage of our approach is the exploitation of DBpedia to supplement the information of the documents. This completion corresponds to:

- geolocation and weather data for destinations,
- essentially a treatment of the possible misinterpretations in the language and country for films.

Apart from completion, the main difference between the use of data by the classifiers and SAUPODOC is the way of representing the texts:

- For classifiers, the use of a bag-of-words with TF-IDF provides a frequency notion that is not in the ontology. Indeed, in the ontology, the presence or absence of a property assertion is a binary notion, much less fine than the TF-IDF.
- For SAUPODOC, the advantage of using an ontology compared to a bag-of-words is the structure. In a bag-of-words, there is no notion of proximity between words, unlike in an ontology where similar individuals are instances of common classes, and similar classes are subclasses of common super-classes.

Figures 7 and 8 show the results on the four measures for the three approaches as well as the SAUPODOC approach without completion of the assertions with data from DBpedia (without Task 2).

For destinations, without completion of geolocation and weather data, cf. Fig. 7, SAUPODOC is less efficient, as it can be expected. However, it keeps surpassing the two classifiers on the four measures. For example, for a target concept where we intuitively think of a nice weather in winter, like *DWW*, SAUPODOC without completion cannot use weather data, since this data is obtained during completion task. Nevertheless, it is able to get a definition partly based on the environment of this type of destinations,

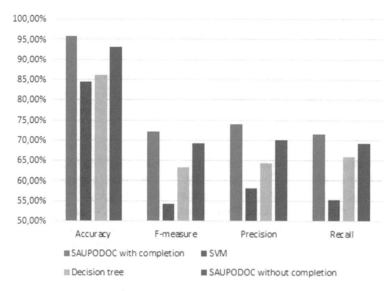

Fig. 7 Results on the corpus of destinations (carried out on the test set)

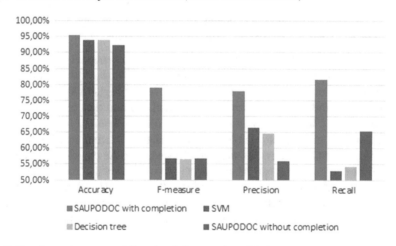

Fig. 8 Results on the corpus of films (carried out on the validation set)

e.g., (hasEnvironment some (Jungle or Vegetation)). Indeed, descriptions of destinations where the weather is good in winter often mention a jungle nearby or at least some vegetation that do not mention the other destinations. Thanks to the structure of the ontology, individuals that are instances of the class Jungle (or of the class Vegetation) are automatically seen as close individuals, unlike in the classifiers. Hence, the structure of the ontology enables to find more correct labels with SAUPODOC without Task 2 than with the classifiers in the case of touristic destinations.

Table 2 Contribution of completion in SAUPODOC. The delta is the difference between the given measure for SAUPODOC with and without completion

Corpus	Delta accuracy (%)	Delta F-measure (%)	Delta precision (%)	Delta recall (%)
Destination	2.82	2.97	3.80	2.26
Film	2.98	22.15	22.08	16.00

In the case of films, without completion of languages and countries, the four measures are considerably lower (cf. Fig. 8). Accuracy is a bit worse than for classifiers. The precision of SAUPODOC without Task 2 is clearly lower while the recall is better, creating roughly the same F-measure as classifiers. Overall, the performance of SAUPODOC without completion is very close to the one of classifiers but a little worse. This is due to the fact that the ontology is weakly structured. It connects the films to 5 concepts, but has no internal structure of concepts, and has only 18 initial individuals. Therefore, the power of ontology, residing in its ability to combine similar individuals by associating them with the same classes, is not really present here. Thus, SAUPODOC can benefit from neither the structure of the ontology, nor the notion of frequency that classifiers benefit from thanks to the bag-of-words TF-IDF.

From these experiments, we can deduce two conclusions:

- Completion has an interest for the approach. Indeed, it generates a certain contribution. For destinations, the added numerical values allow a gain between 2 and 4% in our measures, see Table 2. For movies, given the higher risk of misinterpretation in textual documents, taking into account external data allows for a large gain: around 3% for accuracy and 20% for the other measures.
- The advantage of using an ontology compared to a conventional bag-of-words resides in the exploitation of its structure. Indeed, here the bag-of-words contains notions of similarity (several labels for the same individual). It is a kind of "bag of extracted relations". However, it ignores the concepts linking the extracted relations between themselves, unlike an ontology. Thereby, the structure of the ontology allows an important contribution in the learning task. Table 3 shows

Table 3 Contribution of the structure of the ontology: case of the destinations. The delta is the difference between the given measure for SAUPODOC without completion and the classifiers

Corpus of destinations	Delta accuracy (%)	Delta F-measure (%)	Delta precision (%)	Delta precision (%)
Delta SAUPODOC without Task 2 compared to SVM	8.56	15.11	12.04	14.00
Delta SAUPODOC without Task 2 compared to decision tree	6.85	6.04	5.91	3.42
Average delta	7.70	10.58	8.98	8.71

Table 4 Absence of contribution without structure of the ontology: case of the films. The delta is the difference between the given measure for SAUPODOC without completion and the classifiers

Corpus of films	Delta accuracy (%)	Delta F-measure (%)	Delta precision (%)	Delta recall (%)
Delta SAUPODOC without Task 2 compared to SVM	−1.52	−0.02	−10.58	12.74
Delta SAUPODOC without Task 2 compared to decision tree	−1.45	0.30	−8.66	11.31
Average delta	−1.49	0.14	−9.62	12.03

that with the same information (only from documents) represented in a bag-of-words or in a well-structured ontology, using an ontology allows a gain in the labeling approach. On the contrary, without structure in the ontology, see Table 4, the ontological approach has no interest in the labeling.

6 Conclusion and Future Works

We have proposed an original approach for automatically labelling of documents describing entities with specific concepts not explicitly mentioned in documents. This approach combines steps of ontology population and enrichment. Tasks cooperate via an ontology under the closed-world assumption, which is quite original for an ontology-based approach. It implements also innovative mechanisms to exploit the LOD without being penalized by its incompleteness.

Experiments show the relevance of such a combined approach. SAUPODOC performs better than classifiers thanks to the structure of the ontology that highlights the semantic links between different individuals, and thanks to completion of information from external resources.

Future work will address experiments on new domains, such as books and music. We are also interested in a semi-automatic refinement of the definitions. Indeed, when the definition of a target concept only generates negative labels, the Wepingo company is not able to propose any entity that fit the user need corresponding to this target concept. The definition could be refined a bit in order to have some positive labels. To do this in a semi-automatic way, we could use the structure of the ontology, for example, by replacing a concept in a definition by one of its super-concept. New definitions generating positive labels would then be proposed to the designer to be validated.

Acknowledgements We acknowledge the Wepingo startup, which has funded this work in the settings of the PORASO project.

References

Alec, C., Reynaud-Delaître, C., & Safar, B. (2016). A model for linked open data acquisition and SPARQL query generation. In *Graph-based Modeling of Conceptual Structures. 22nd International Conference on Conceptual Structures, ICCS* (pp. 237–251). Annecy, France: Springer.

Bontcheva, K., Tablan, V., Maynard, D., & Cunningham, H. (2004). Evolving GATE to meet new challenges in language engineering. *Natural Language Engineering, 10*(3/4), 349–373.

Cheng, X., & Roth, D. (2013). Relational inference for wikification. *Empirical Methods in Natural Language Processing (EMNLP)* (pp. 1787–1796), Seattle, Washington, USA.

Chitsaz, M. (2013). Enriching ontologies through data. In *Doctoral Consortium Co-located with International Semantic Web Conference (ISWC)* (pp. 1–8), Sydney, Australia.

Cimiano, P. (2006). *Ontology learning and population from text: Algorithms. Evaluation and applications.* Secaucus, NJ, USA: Springer New York Inc.

Cimiano, P., & Völker, J. (2005). Text2Onto: A framework for ontology learning and data-driven change discovery. In *Proceedings of the 10th International Conference on Natural Language Processing and Information Systems, NLDB* (pp. 227–238). Alicante, Spain: Springer.

Cimiano, P., Völker, J., & Studer, R. (2006). Ontologies on demand?–A description of the state-of-the-art, applications, challenges and trends for ontology learning from text. *Information, Wissenschaft und Praxis, 57*(6–7), 315–320.

Cunningham, H., Maynard, D., Bontcheva, K., Tablan, V., Aswani, N., Roberts, I., Gorrell, G., Funk, A., Roberts, A., Damljanovic, D., Heitz, T., Greenwood, M. A., Saggion, H., Petrak, J., Li, Y., & Peters, W. (2011). *Text Processing with GATE.* ACM Digital Library.

Esposito, F., Fanizzi, N., Iannone, L., Palmisano, I., & Semeraro, G. (2004). Knowledge-intensive induction of terminologies from metadata. In *Third International Semantic Web Conference (ISWC), Hiroshima, Japan, November 7–11* (pp. 441–455).

Fanizzi, N., d'Amato, C., & Esposito, F. (2008). DL-FOIL concept learning in description logics. *18th International Conference Inductive Logic Programming, (ILP)* (pp. 107–121). Prague, Czech Republic.

Lehmann, J. (2009). DL-Learner: Learning concepts in description logics. *Journal of Machine Learning Research, 10,* 2639–2642.

Lehmann, J., Auer, S., Bühmann, L., & Tramp, S. (2011). Class expression learning for ontology engineering. *Journal of Web Semantics, 9,* 71–81.

Lehmann, J., & Hitzler, P. (2010). Concept learning in description logics using refinement operators. *Machine Learning, 78*(1–2), 203–250.

Ma, Y., & Distel, F. (2013a). Concept adjustment for description logics. *7th International Conference on Knowledge Capture, K-CAP'13* (pp. 65–72). Banff, Canada: ACM.

Ma, Y., & Distel, F. (2013b). Learning formal definitions for snomed CT from text. In *Proceedings of Artificial Intelligence in Medicine (AIME)* (pp. 73–77). Murcia, Spain: Springer.

Mendes, P. N., Jakob, M., García-Silva, A., & Bizer, C. (2011). DBpedia spotlight: Shedding light on the web of documents. *7th International Conference on Semantic Systems, I-Semantics'11* (pp. 1–8). NY, USA: ACM.

Petasis, G., Möller, R., & Karkaletsis, V. (2013). BOEMIE: Reasoning-based information extraction. *12th International Conference on Logic Programming and Nonmonotonic Reasoning (LPNMR)* (pp. 60–75), A Corunna, Spain.

Ratinov, L., Roth, D., Downey, D., & Anderson, M. (2011). Local and global algorithms for disambiguation to wikipedia. In *49th Annual Meeting of the Association for Computational Linguistics (ACL)* (pp. 1375–1384).

Shearer, R., Motik, B., & Horrocks, I. (2008). HermiT: A highly-efficient OWL reasoner. In *Fifth Workshop on OWL (OWLED), Co-located with the 7th International Semantic Web Conference,* volume 432 of *CEUR Workshop Proceedings.*

Sirin, E., Parsia, B., Grau, B. C., Kalyanpur, A., & Katz, Y. (2007). Pellet: A practical OWL-DL reasoner. *Journal of Web Semantics, 5*(2), 51–53.

Tsarkov, D., & Horrocks, I. (2006). FaCT++ description logic reasoner: System description. In *Third International Joint Conference Automated Reasoning (IJCAR)* (pp. 292–297), Seattle, WA, USA.

Völker, J., Hitzler, P., & Cimiano, P. (2007). Acquisition of OWL DL axioms from lexical resources. In *4th European Semantic Web Conference (ESWC)*, pp. 670–685. Innsbruck, Austria: Springer.

Yelagina, N., & Panteleyev, M. (2014). Deriving of thematic facts from unstructured texts and background knowledge. *5th International Conference Knowledge Engineering and the Semantic Web (KESW)* (pp. 208–218). Kazan, Russia: Springer.

Yosef, M. A., Hoffart, J., Bordino, I., Spaniol, M., & Weikum, G. (2011). AIDA: An online tool for accurate disambiguation of named entities in text and tables. In *Proceedings of the 37th International Conference on Very Large Databases, (VLDB)* (pp. 1450–1453).

Author Biographies

Céline Alec obtained her PhD in 2016 from the Université Paris-Sud, Paris-Saclay, France. Her supervisors were Pr. Chantal Reynaud-Delaître and Dr. Brigitte Safar, members of the LaHDAK team (Large-scale Heterogenous DAta and Knowledge) in the LRI (Laboratoire de Recherche en Informatique). Her research interests are situated within the domain of ontology enrichment and population applied to semantic annotation of documents.

Chantal Reynaud-Delaître is a professor of Computer Science in the Laboratory for Computer Science (LRI) at the University of Paris-Sud. For several years, she was the head of the LaH-DAK (Large-scale Heterogeneous Data and Knowledge) team. Her areas of research are Ontology Engineering and Information Integration. In particular, she works on the following topics: integration of semantically heterogeneous information sources, linked open data integration and maintenance of semantic annotations impacted by the evolution of ontologies. She is involved in several projects, as the Labex DigiCosme at the University Paris-Saclay and is the author of more than 120 refereed journal articles and conference papers.

Brigitte Safara is an Assistant Professor of Computer Science in the Laboratory for Computer Science (LRI) at the University of Paris-Sud. She is a member of the LaHDAK (Large-scale Heterogeneous Data and Knowledge) team. Her research interests are Ontology Engineering and Information Integration.

C-SPARQL Extension for Sampling RDF Graphs Streams

Amadou Fall Dia, Zakia Kazi-Aoul, Aliou Boly and Yousra Chabchoub

Abstract Our daily use of Internet and related technologies generates continuously large amount of heterogeneous data flows. Several RDF Stream Processing (RSP) systems have been proposed. Existing RSP systems benefit from the advantages of semantic web technologies and traditional data flow management systems. C-SPARQL, CQELS, SPARQL$_{stream}$, EP-SPARQL, and Sparkwave extend the semantic query language SPARQL and are examples of those systems. Considering that the storage and processing of all these streams become expensive, we propose a solution to reduce the load while keeping data semantics, and optimizing treatments. In this paper, we propose to extend C-SPARQL for continuously generating samples on RDF graphs. We add three sampling operators (UNIFORM, RESERVOIR and CHAIN) to the C-SPARQL query syntax. These operators have been implemented into Esper, the C-SPARQL's data flow management module. The experiments show the performance of our extension in terms of execution time and preserving data semantics.

1 Introduction

Today we produce more data than resources to process them. Our daily use of social networks (Facebook, Twitter, Linkedin, etc.), contents of social multimedia platforms (YouTube, Flickr, iTunes, etc.), sensor networks (observation, remote reading, etc.), internet of things (geolocation, triggering real-time alarms, etc.), etc. produces

A.F. Dia (✉) · Z. Kazi-Aoul · Y. Chabchoub
ISEP, 75006 Paris, France
e-mail: amadou.dia@isep.fr

Z. Kazi-Aoul
e-mail: zakia.kazi@isep.fr

Y. Chabchoub
e-mail: yousra.chabchoub@isep.fr

A. Boly
Cheikh Anta Diop University, BP 5005 Dakar-fann, Senegal
e-mail: aliou.boly@ucad.edu.sn

© Springer International Publishing AG 2018
B. Pinaud et al. (eds.), *Advances in Knowledge Discovery and Management*,
Studies in Computational Intelligence 732, https://doi.org/10.1007/978-3-319-65406-5_2

continuous data streams. Several research groups are interested in semantic web technologies application to real time stream processing. Like DSMSs (Data Stream Management Systems), several extensions to SPARQL have been proposed for processing RDF streams. The six (6) major propositions languages and/or sytems are Streaming SPARQL (Bolles et al. 2008), Continuous SPARQL (C-SPARQL) (Barbieri et al. 2010), CQELS (Le-Phuoc et al. 2011), SPARQL$_{stream}$ (Calbimonte et al. 2010), EP-SPARQL (Anicic et al. 2011a) and Sparkwave (Komazec et al. 2012). They all extend SPARQL but adopt different approaches to deal with continuous RDF streams.

As the volume and the unpredictable speed of incoming data increase, processing the entire contents of a stream is difficult. Systems need therefore techniques for (i) dynamic resources allocation (Vijayakumar et al. 2010) and (Cao et al. 2012) or (ii) reduction of the input data load. Regarding this last point, when input stream rate is high (exceeds capabilities of RSP engines), systems will be overloaded. For instance, a high query execution time compared to the input stream rate will cause overload and thus loss of important data and latency in processing. To keep pace of data arrival, the system would shed some of the load according to a given method. None of existing SPARQL extensions implement continuous data summaries or sampling mechanisms.

In this paper, we propose to extend C-SPARQL engine by adding sampling operators. These operators will allow us to reduce, on the fly, input RDF data, while preserving semantic links. We consider first a new data input format. Indeed, C-SPARQL takes as input a sequence of pairs, where each pair is made of an RDF triple and its timestamp. This form of representation does not guarantee the semantic links between data within the sample. In the context of RDF data streaming, events are frequently captured by a set of triples but not by only one triple. Thus, instead of RDF triple format (*<subject, predicate, object>, timestamp*), we adopt a graph oriented approach where each graph represents an event formed by a set of temporal RDF triples. We then extend the C-SPARQL query syntax and the continuous execution module of its architecture (Esper) by adding new sampling operators.

This paper is organized as follows. Section 2 introduces RDF graphs streams concept by deducting the triples format. We provide a brief state of the art on existing RDF Stream Processing (RSP) systems in Sect. 3. Section 4 gives the three sampling algorithms used for the implementation of our sampling operators while we present in Sect. 5 the extended C-SPARQL syntax and architecture. Section 6 presents our performance evaluation and results. Finally, we conclude and give our future works in Sect. 7.

2   RDF Graphs Streams

In this section, we present the notion of RDF graph streams adopted for sampling process in C-SPARQL.

2.1 RDF Streams: Triple Based

RDF (Resource Description Framework) is the formal W3C recommendation for semantic data representation. The base element of the RDF model is the triple: $<s, p, o> \in (I \cup B) \times I \times (I \cup B \cup L)$ where I is a set of IRIs (Internationalized Resource Identifiers), B is a set of blank nodes and L is a set of literals. s, p, and o represent information and are respectively called the subject, the predicate and the object of the RDF triple. However, RDF model lacks of temporal dimension which is compulsory in the context of streaming. Thus, applying semantic web technologies on streaming data has given rise to the notion of RDF streams.

Golab and Özsu (2003) define a data stream as follow: "*a data stream is a real-time, continuous, ordered (implicitly by arrival time or explicitly by timestamp) sequence of items. It is impossible to control the order in which items arrive, nor is it feasible to locally store a stream in its entirety*". Then RDF Streams are introduced as natural extension of RDF model in streaming context. An RDF stream S can be defined as a temporal ordered sequence of pairs, where each pair consists of an RDF triple and its timestamp: $S = \langle < t_r, \tau_i > \rangle$, where t_r is a triple observed or arrived at time τ_i. Integers τ_i are monotonically non-decreasing and not strictly increasing ($\forall i, \tau_i \leq \tau_{i+1}$).

$$\langle < subject_i, predicate_i, object_i >, \tau_i \rangle$$

$$\langle < subject_{i+1}, predicate_{i+1}, object_{i+1} >, \tau_{i+1} \rangle$$

$$...$$

Given that streams are intrinsically infinite, data are usually read through windows upon streams using the CQL (Arasu et al. 2004b) window concept. Queries over RDF streams deal with triples in time-based window (all the triples which occur during a given time interval) or element-based window (a given number of triples).

Most of the existing SPARQL extensions for streaming RDF streams take as input a succession of RDF triples. This representation model allows widely continuous process of streams but consumes only triples, ignoring the graph structure of RDF data. In this case, each event is distributed into a set of triples. Considering stream of successive triples, we can not capture boundaries on a set of triples within different events. Therefore, this succession of triples which belongs to a same event needs to be grouped into a single graph and annotated with the same timestamp.

2.2 RDF Streams: Graph Based

As preliminaries of our work, we consider an RDF graph stream format by extending the definition of RDF stream format for this purpose. Events within streams are naturally captured by a set of triples with the same timestamp or not. Hence, a graph represents those triples with the same source and unique timestamp.

An RDF graph may be constructed from one or a set of RDF triples or statements. More formally, given a set of triples st_r, we define $G\langle st_r\rangle$ as a directed labelled graph where each vertex consists of triple's subject (s) or object (o) and each edge consists of triple's predicate (p). The notion of connectivity in RDF graph definition is important for all triples which compute an event. In fact, triples within an event share between each other a unique or multipath. A connectivity in RDF graph $G\langle st_r\rangle$ is a sequence of edges (predicates) $p_1, ..., p_n$ such as $\forall 1 \leq i < n$, p_i share a vertex (subject or object) with p_{i+1}.

With the definition above, RDF graph stream can be simply defined as a sequence of pair $(G(st_r)_i, \tau_i)$, where $G\langle st_r\rangle_i$ is an RDF graph represented as an event and τ_i is the time when this event occurs.

$$(G\langle st_r\rangle_i, \tau_i)$$

$$(G\langle st_r\rangle_{i+1}, \tau_{i+1})$$

...

Two graphs may share a same timestamp value, which means that their events occurs at the same time. Highlight that two triples with the same timestamp are not necessary within the same event. Figure 1 shows an example of a flow of RDF graphs from sensors deployed in a water distribution network. The data collected concern the pressure, the flow rate, the chlorine content, the temperature, etc.

In particular, sampling RDF streams when events are decomposed into a list of individual RDF triples may break semantic links between them in the sample. In that case, samples may end up with meaningless data and queries over them only observe a partial RDF graph and obviously return false or incomplete results. Therefore, we adopt a graph-oriented approach that ensures the preservation of data semantic (i.e. meaning links between subjects and objects) after a sampling operation.

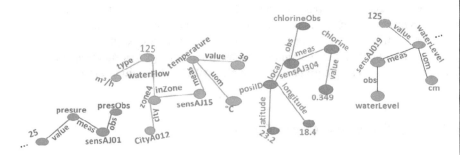

Fig. 1 Example of graph streams from water distribution network

3   RDF Stream Processing Systems

All proposed SPARQL extensions for stream processing share at the basis the same issues in terms of heterogeneity (multiple data sources), and lack of explicit data semantic allowing complex queries and reasoning over data. Several systems exist such as Streaming SPARQL, C-SPARQL, CQELS, SPARQL$_{stream}$, EP-SPARQL and Sparkwave. In this section, we give a brief state of the art of those SPARQL extensions.

Streaming SPARQL (Bolles et al. 2008) extends SPARQL grammar with the capability to explicitly state data streams and define physical or logical windows over them. Authors modify the semantics of SPARQL by first adding the STREAM keyword to allow as input a data stream identified by an IRI. Then, they consider two window types. The time based window is defined with the RANGE (a definition of the window size), SLIDE (a delay after which window is moved) and FIXED (in case which RANGE value equals SLIDE value). The element based window is defined with ELEMS (a number of given elements) keyword. Authors focus only on an extension to cope with window queries over data streams without taking into account performance issues. Since they extend SPARQL 1.0 version, they therefore do not support group by clauses and aggregation which are usually necessary in streaming context. There is no proposed Streaming SPARQL system. The following Streaming SPARQL query returns every minute, pressure values captured in the last 10 min.

SELECT   ?WaterPressure
FROM STREAM <http://waterdist.org/sens> **WINDOW RANGE** 10 MINUTE
                                                                              SLIDE
WHERE { ?sensorID ex:hasPressure ?WaterPressure.}

Continuous SPARQL or shortly **C-SPARQL** (Barbieri et al. 2010) proposes a continuous query syntax and a framework which reuses independently existing and tested technologies such as data stream management systems like STREAM (Arasu et al. 2004a) or ESPER[1] and triple stores Jena[2] or Sesame.[3] Authors also propose their own operators for windowing (inspired from CQL, the query language used in relational stream systems) and aggregation functions. C-SPARQL also allows periodical query evaluation, temporal function and multiple streams. Authors first extends SPARQL 1.0 but switched to support SPARQL 1.1 in their latest version to benefit from new functions like aggregation ones. The following C-SPARQL query example returns every 10 min, the average pressure values captured in the last hour.

REGISTER QUERY AverageWaterPressure **AS**
SELECT   ?sensorID   (AVG(?WaterPressure) as ?AvgWaterPressure)

---

[1]http://www.espertech.com/esper/.
[2]https://jena.apache.org/.
[3]http://rdf4j.org/sesame/.

FROM STREAM <http://waterdist.org/sens> [**RANGE** 1h **STEP** 10m]
WHERE { ?sensorID ex:hasPressure ?WaterPressure.}
GROUP BY   ?sensorID

Like C-SPARQL, **CQELS** (Le-Phuoc et al. 2011) language adds its own window operators analogous to CQL and allows possibilities to combine static and streaming data processing. Unlike C-SPARQL, CQELS applies an eager execution strategy which means that query evaluation is triggered by the arrival of new triples. The system does not delegate stream evaluation to an external component (DSMS) but defines its own native processing model, which is implemented internally in the query engine. Author's strategies allow full control over the execution plan consequently, enabling query optimisations. The following CQELS query retrieves every 5 min, the minimal value of pressure captured in the latest hour.

SELECT   ?sensorID   (MIN(?WaterPressure) as ?MinWaterPressure)
FROM STREAM <http://waterdist.org/sens> [**RANGE** 1h **SLIDE** 5m]
WHERE { ?sensorID ex:hasPressure ?WaterPressure.}
GROUP BY   ?sensorID

$SPARQL_{stream}$'s authors (Calbimonte et al. 2010) were at base inspired by C-SPARQL systems. Similarly to it, $SPARQL_{stream}$ delegates the processing to external engines. $SPARQL_{stream}$ also proposes its own stream and window operators and aims more at enabling ontology-based access for semantic stream processing. The proposed system (Morph-Stream) is based on the concept of virtual RDF streams where each of them is identified by an IRI. $SPARQL_{stream}$ only considers time-based window. $SPARQL_{stream}$ is the only proposed extension which supports all Relation-to-Stream operators (R_{stream}, I_{stream} and D_{stream}) inspired from CQL. The following $SPARQL_{stream}$ query shows sensors list having observed, in the last hour, a temperature value above 60 °C.

SELECT   ?sensorID
FROM NAMED STREAM <http://waterdist.org/sens> [NOW-1 HOURS]
WHERE { ?sensorID ex:hasTemperature ?temperature.
                   FILTER (?temperature > 60)}

EP-SPARQL (Anicic et al. 2011a) delegates the processing to an event processing engine. Unlike the previous extensions which focus on windows and stream registering, EP-SPARQL proposes an unified language for event processing. The language follows the concept of Complex Event Processing (CEP) systems. For expressing complex event of input data, EP-SPARQL mainly includes logical and temporal operators expressed in terms of sequence and simultaneity. SEQ operator specifies that two graphs patterns are joined if one occurs after the other. EQUALS operator placed between two patterns, indicates that they are joined if they occur at the same time. OPTIONALSEQ and EQUALSOPTIONAL represent respectively optional version of SEQ and EQUALS operators. EP-SPARQL simulates window operators by using filter operators to isolate portions of the input data streams. While other systems use a single point in time approach to represent each pattern in the stream, EP-SPARQL adopts an interval approach which represents the lower and upper bound of the occurring interval. The system ETALIS (Anicic et al. 2011b) is a complex event processing framework based on Prolog language. The following EP-SPARQL query gives all pressure sensors whose measured values have dropped by more than half and then have dropped by over a quarter in a period of 24 h.

SELECT ?sensorID
WHERE { ?sensorID ex:hasPressure ?waterPressure$_1$ }
          SEQ { ?sensorID ex:hasPressure ?waterPressure$_2$ }
          SEQ { ?company ex:hasPressure ?waterPressure$_3$ }
          FILTER (?waterPressure$_2$ < ?waterPressure$_1$/2 && ?waterPressure$_3$ <
          ?waterPressure$_1$/4 && getDURATION() < "P24H"8sd:duration)

Sparkwave (Komazec et al. 2012) is a complex system which adopts graph pattern detection on RDF data streams and supports temporal nature of RDF streams. Sparkwave implements windowing mechanism based on Rete algorithm (Rete 1982) (with ε-network Pre-processing). Sparkwave supports time-based windows but does not support temporal operators, arithmetic operators and logical operators (disjunctions and negations). In fact, as mentioned above, Sparkwave only supports graph pattern detection which constitutes a limit of the system.

Table 1 briefly compares features and functionalities of SPARQL extensions for stream processing. In general, they all have similar approaches for supporting temporal and window process over streams. They also all extend SPARQL but have small or important differences depending on the field of application.

Streaming SPARQL seems to be the most limited system in terms of operators. It lacks of group by and aggregation operators and does not support SPARQL 1.1. Moreover, the proposed system is still theoretical. **C-SPARQL, CQELS** and **SPARQL$_{stream}$** come with a SPARQL language extension and a query processing engine. They are also based on SPARQL 1.1 and all support sliding windows. Among these three engines, only SPARQL$_{stream}$ lacks of window based elements. **EP-SPARQL** is different from the other proposed extensions as they adopt sequence (SEQ and OPTIONALSEQ) and simultaneity (EQUALS and EQUALSOPTIONAL) operators following Complex Event Processing (CEP) paradigm. Like EP-SPARQL, **Sparkwave** focused on event processing. However, Sparkwave has no SEQ or EQUALS operators but is only based on Rete algorithm for pattern matching. Unlike the other systems, which only deal with algebraic optimisation, CQELS and Sparkwave use native approach and then can bring adaptive optimisations. In C-SPARQL and SPARQL$_{stream}$, the data is stored in relational tables and relational streams before any further processing. In our opinion, among all, C-

Table 1 Comparison of SPARQL extensions for RDF stream processing

System	DIF[a]	TiW[b]	TrW[c]	U, J, O, F[d]	TF[e]	A[f]	S/S[g]	CQ[h]
Streaming SPARQL	RS[i]	✓	✓	✓	✗	✗	✗	✓ (P[l])
C-SPARQL	RS & S[j]	✓	✓	✓	✓	✓	✗	✓ (P[l])
CQELS	RS & S[j]	✓	✓	✓	✗	✓	✗	✓ (T[m])
SPARQL$_{stream}$	(V)RS[k]	✓	✗	✓	✗	✓	✗	✓ (P[l])
EP-SPARQL	RS & S[j]	✗	✗	✓	✓	✓	✓	✓ (P[l])
Sparkwave	RS & S[j]	✓	✗	✓	✗	✗	✗	✓ (T[m])

[a] Data Input Format
[b] Time Window
[c] Triple Window
[d] UNION, JOIN, OPTIONAL, FILTER
[e] Temporal Function
[f] Aggregate
[g] Sequence/Simultaneity
[h] Continuous Query
[i] RDF streams
[j] RDF stream & Statics
[k] Virtual RDF Stream
[l] Periodic
[m] Trigger

SPARQL is the only system that provides at the same time Union, Join, Optional and Filter operators, logical and physical window, aggregation, continuous processing, multiple streams sources, combining static and RDF stream evaluation, temporal function, and is built on a modular architecture.

4 Sampling Algorithms

RDF stream processing systems require rapid, continuous and intelligent data processing. Given the volume and speed of data generation, it is necessary to extract data samples from input streams. Several sampling techniques exist such as random sampling, reservoir sampling, deterministic sampling and chain sampling.

4.1 Uniform Random Sampling Without Replacement

The simple random sampling (Cochran 2007) may be with or without replacement. It selects without replacement and with the same probability p a random sample of size n from a set of indexes within a window W. The index of an element in W can be selected only once.

This method is very basic and has the advantage of being simple and easy to implement. However, this technique gives to all elements the same chance of being included in the sample. This can be seen as a disadvantage because in a streaming context we are often interested in recent data. In the sample constitution, we should give a more chance for relatively recent data.

4.2 Reservoir Sampling

The main idea of any type of reservoir sampling algorithm (Vitter 1985) is maintaining a random sample with a fixed size n into a "reservoir". After each windowing process over streams, a random sample of size n can be extracted. Initially, we put the first n received items into the "reservoir" R. Then, each new item in the window has the probability $\frac{n}{i}$ to replace the item of index i in the reservoir R. This method clearly favors old items ($\lim_{i \to \infty} \frac{n}{i} = 0$). Therefore, oldest items are more likely to be included in the sample.

4.3 Chain Sampling

Babcock et al. (2002) described that the chain sampling technique consists in building a sample of size n over a sliding windows of size $\omega > n$. For sliding windows, chain sampling algorithm randomly generates replacements among expired items, and stores the replacement. As shown in chain sampling algorithm, in the first window, we add i indexes in the sample with the probability $p = \frac{min(\omega,i)}{\omega}$. The successor index r_i of index i is randomly chosen from the interval $[i + 1, i + \omega]$ and replaces it in the sample after its expiration (i out of the window). The successor of r is randomly selected in the same manner in the interval $[r + 1, r + \omega]$. This process is thus repeated independently.

Algorithm 1: Chain sampling algorithm

---

1 **Function** chainSamp (ω, p):
2     $R_{epl} \leftarrow \varnothing$
3     $S \leftarrow$ put indexes (i) from the first window (with size ω) with probability $\frac{Min(\omega,i)}{\omega}$
4   **for** *each index i in S* **do**
5   Select a random successor r_i with probability p between $i + 1$ and $j + \omega$
6         $R_{epl} \leftarrow r_i$
7   **while** *each new index is added* **do**
8         $i \leftarrow i + 1$ //*move window index by on step*
9   Replace each expired index j in S by its successor r_j in R without redondancy
10   Choose a random successor for r_j between $r_j + 1$ and $r_j + \omega$ without redondancy
11   **return** S

This method is particularly suitable for sliding windows but has a significant memory usage due to the non-redundant selection criterion for successors.

5 Our C-SPARQL Extension

C-SPARQL system (Barbieri et al. 2010) provides a modular architecture for processing C-SPARQL queries over RDF streams. This architecture is composed of two parts: a data stream management system like STREAM or Esper and a SPARQL engine module like Jena or Sesame.

This architecture uses STREAM or Esper as RDF streams manager module and Jena or Sesame as SPARQL engine module.

Figure 2 presents the C-SPARQL architecture. The *Query Translator* module configures, initializes and dispatches continuous and static parts of a C-SPARQL query. This dispatching task processes a correct C-SPARQL query and creates two (2) instances:

1. ***ContinuousEngine*** consists of a DSMS (Esper in the last version) that processes on the fly RDF triples and applies the concept of window over them through a CQL query. This module outputs a set of quadruplets (*subject, predicate, object, timestamp*) intended to the SPARQL engine SparqlEngine.

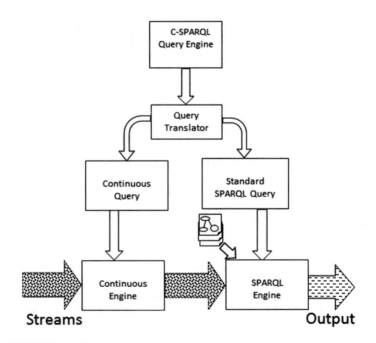

Fig. 2 C-SPARQL architecture

2. *SparqlEngine* is the "semantic" component in this all architecture. Its main task is running SPARQL deduced from the C-SPARQL one, each time the ContinuousEngine produces quadruples.

The main contribution of our work is the extension of the C-SPARQL syntax and architecture level for continuous sampling of RDF graphs. To reduce the load of data streams to process and store only a sample for potential analysis or reasoning, our implementation includes three (3) main steps:

1. Considering input graphs data streams instead of an ordered sequence of RDF triples for semantic preservation
2. Adding three (3) sampling operators within C-SPARQL query syntax
3. Implementing continuous sampling methods within Esper.

5.1 Sampling Operators

As shown below, we extend the C-SPARQL query syntax by adding new operators in **FROM STREAM** (stream sources) and **FROM** (static sources) clauses.

---

1. *PREFIX *prefixName* : < *IRI* >
2. SELECT '*?variables*'
3. *FROM STREAM < *StreamIRI* > [*Window*] **[SAMPLING   Token]**
4. *FROM < *StaticIRI* > **[SAMPLING   Token]**
5. WHERE { '*Mapping variables*' ; |.
6.   *FILTER ('*condition*')}
7. GROUP BY '*?variables*'|*expression*
8. HAVING '*aggregation condition*'
9. ORDER BY '*?variables*'

SAMPLING → *UNIFORM* | *RESERVOIR* | *CHAIN*
Token → [*Window*]   %*P*|*Size*
    [*Window*] → '*chain window size*'
    %*P* → '*sampling percentage*'
    Size → '*reservoir size*'

The extended syntax contains new operators for sampling methods continuously applied over input RDF graph streams and statics (RDF repositories). The following query randomly samples input graphs and provides every minute, all sensors whose sensed pressure value exceeds 27 (unit of measure) in the last 10 min.

REGISTER QUERY exceedsPressures AS
SELECT   ?sensorID   ?WaterPressure
FROM STREAM <http://waterdist.org/sens> [RANGE 10m STEP 1m]
                                              [UNIFORM %60]
FROM <http://waterdistrib.org/staticdata.rdf>
WHERE {?sensorID ex:hasPressure ?WaterPressure.
    FILTER (?WaterPressure > 27)
        }

5.2   Architecture Extension

Our approach is based on the implementation of sampling methods within Esper.
Figure 3 shows our proposed extension of C-SPARQL architecture where its tradi-
tional modules remain independent plugins. We extend the three (3) modules *Query
Translator*, *ContinuousEngine* and *SparqlEngine*.

1. In **QueryTranslator**, we parse the received request by first checking sampling
   operators contained in *FROM* and *FROM STREAM* lines. If the request does not
   include a sampling operator, input RDF streams are processed continuously without any sampling phase. If not, after validation we create two (2) instances both

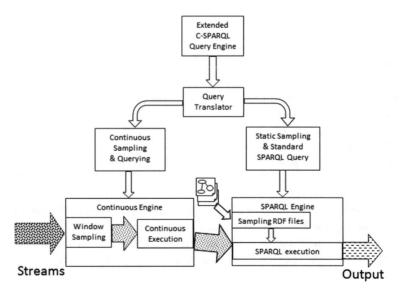

Fig. 3   C-SPARQL extended architecture

ContinuousSampQuery and *SparqlSampQuery* that will be processed respectively by *ContinuousEngine* and *SparqlEngine*.

2. **ContinuousEngine** module receives a continuous query with associated sampling operators. Each sampling method is continuously applied on a window of input graph streams (*WindowSampling*). We then process the CQL part of C-SPARQL query on graphs within our buffered sample according to Esper syntax. Results are transmitted to the third module *SPARQLEngine*.

3. **SPARQLEngine** module also receives graphs from sampled static RDF repository (*SampStaticRDF*). It runs, in last step, the SPARQL query over continuous and static graphs already sampled.

6   Evaluation

This section evaluates the quality and relevance of our extension. To do this, we focus on performance achieved in terms of execution time and preserving data semantics. As a case study, we consider a set of data from sensors deployed in a large water distribution network. Data observed can be pressure, flow, chlorine or temperature. The management of water distribution network needs capabilities for real time processing and reasoning over streams in order to rapidly detect anomalies like water leaks. We consider the treatment of 80000 graphs, where each graph contains 10 triples. This set of data is sent in form of graphs and triples streams respectively at rates of 500 graphs/s and 5000 triples/s. Experiments in this section are performed on a computer with a processor at 2.66 GHz, Core 2 Duo and 4 GB of RAM.

For performance evaluation of query execution time, we consider the simple query given below. The aim of the query is to return the average pressure values captured by each sensor. The query is performed over 1000 sampled graphs with respectively UNIFORM, RESERVOIR and CHAIN operators.

REGISTER QUERY AvgWaterPressure AS
PREFIX ex: <http://waterdist.org/>
SELECT   ?sensorID   (AVG(?pressureValue) AS ?AvgPressure)
FROM STREAM <http://waterdist.org/stream> [RANGE TRIPLES 1000]
                    [SAMPLING [window] percent|size]
WHERE { ?sensorID ex:hasPressure ?pressureValue. }
GROUP BY   ?sensorID

For sampling operators (*UNIFORM*) and (*RESERVOIR*) Figs. 4 and 5, the evolution of the query processing time by respectively varying the sampling percentage (*percent*) and the reservoir size (*size*). Finally, with CHAIN operator we evaluate in Fig. 6, the query execution time under two conditions: varying sampling percentage and Chain window size (*Window*).

Fig. 4 Uniform

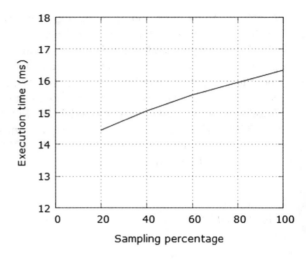

Fig. 5 Reservoir

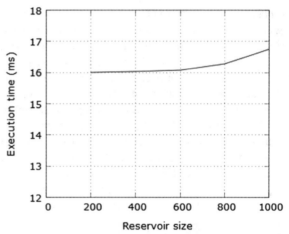

With uniform random sampling without replacement (Fig. 4), the execution time increases depending on the sampling percentage. We also note a similar trend with reservoir sampling (Fig. 5) depending on reservoir size kept fixed in memory. The evolution of the execution time with chain sampling (Fig. 6) depends on the window size and the ratio. Whatever the sampling percentage, query execution time follows a growing trend. This can be explained by the selection of a random successor items without redundancy. Thus, observations confirm performances gained in reducing on the fly the load from input streams.

For the evaluation of the semantic links preservation between data in sample, we consider the following query:

Fig. 6 Chain

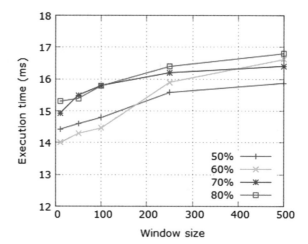

```
REGISTER QUERY sourceSensorID AS
PREFIX ex: <http://waterdist.org/>
SELECT    ?sensorID    ?pressureValue
FROM STREAM <http://waterdist.org/stream> [RANGE TRIPLES 10]
                    [SAMPlING [window] percent|size]
WHERE { ?sensorID ex:hasPressure ?pressureMnemonic.
        OPTIONAL{ ?pressureMnemonic ex:value ?pressureValue.} }
ORDER BY    ?sensorID
```

We first execute this query continuously, in base scenario (without any sampling operation), and then with triple-oriented sampling and finally with graph-oriented sampling. The request selects for the last 10 graphs or triples observed in sample, the sensor ID and its corresponding pressure value. In the graph, the sensor identifier is connected to its captured pressure value through an another node. We compute for each sampling method (triple-based and graph-based), the number of correct and complete results (e.g. sensor ID and corresponding pressure value). We subsequently calculate the loss rate in both cases (triple-oriented and graph-oriented) using the following formula:

$$\textbf{Loss rate}(\%) = \frac{\textbf{NbrNS} - \textbf{NbrWS}}{\textbf{NbrNS}} * 100, \ where \tag{1}$$

NbrNS = Number of correct and complete results in base scenario
NbrWS = Number of correct and complete results with sampling.

Table 2 shows the loss rates calculated by varying sampling parameters of uniform and reservoir methods. The number of correct and complete results in base scenario is **NbrNS = 10625**.

Table 2 Loss rate between triple-oriented and graph-oriented sampling

Operator	Triple oriented sampling		Graph oriented sampling	
	Correct and complete results	**Loss rate** (%)	Correct and complete results	**Loss rate** (%)
UNIFORM				
P = 20%	19	**99.82**	2108	**80.16**
P = 40%	73	**99.31**	3614	**65.98**
P = 80%	181	**98.2**	7137	**32.82**
RESERVOIR				
Size = 2	24	**99.77**	2117	**80.07**
Size = 4	144	**98.64**	4210	**60.37**
Size = 8	522	**95.08**	7240	**31.85**

With graph-oriented sampling, we observe a lower than **loss rate** with triple-oriented. Indeed, this can be explained by random sampling in the case of the triple-oriented where the triple containing a sensor identifier and the one containing its corresponding pressure value might not be both in the sample. In comparison, graph-oriented sampling maintains semantic links ensuring that sensors are always associated with their corresponding pressure values in the sample.

7 Conclusion and Future Works

Massive data stream management is a permanent industrial concern and a scientific challenge. The application of semantic web technologies for data stream remains troubled by the current volumes and rapid generation of streams. In this paper, we took advantages of existing sampling techniques and proposed an extension of the C-SPARQL system for real-time sampling of RDF graph streams. The oriented-graph approach allowed us to preserve semantic links within samples and thereby improved its representativeness. We implemented three sampling operators with different performances observed in terms of execution time.

We are interested in our future work into sampling on the fly over RDF streams using methods based on biased algorithm by associating different weights to the semantic stream items. We will also take into account specificities and background (context-based) to produce suitable samples, in a distributed environment.

Acknowledgements This work was performed under the FUI Waves project. This project aims to design and develop a distributed processing platform of massive data streams. The case study concerns the real-time monitoring of a drinking water distribution network.

References

Anicic, D., Fodor, P., Rudolph, S., & Stojanovic, N. (2011a). Ep-sparql: A unified language for event processing and stream reasoning. In *Proceedings of the 20th international conference on World wide web* (pp.635–644). ACM.
Anicic, D., Fodor, P., Rudolph, S., Stuhmer, R., Stojanovic, N., & Studer, R. (2011b). Etalis: Rule-based reasoning in event processing. *Reasoning in Event-Based Distributed Systems, 347*, 99.
Arasu, A., Babcock, B., Babu, S., Cieslewicz, J., Datar, M., Ito, K., et al., (2004a). Stream: The stanford data stream management system. Book chapter.
Arasu, A., Babu, S., & Widom, J. (2004b). Cql: A language for continuous queries over streams and relations. In *Database Programming Languages* (pp. 1–19). Springer.
Babcock, B., Datar, M., & Motwani, R. (2002). Sampling from a moving window over streaming data. In *Proceedings of the thirteenth annual ACM-SIAM symposium on Discrete algorithms* (pp. 633–634). Society for Industrial and Applied Mathematics.
Barbieri, D. F., Braga, D., Ceri, S., & Grossniklaus, M. (2010). An execution environment for c-sparql queries. In *Proceedings of the 13th International Conference on Extending Database Technology* (pp. 441–452). ACM.
Bolles, A., Grawunder, M., & Jacobi, J. (2008). Streaming sparql-extending sparql to process data streams. *The semantic web: research and applications* (pp. 448–462)
Calbimonte, J.-P., Corcho, O., & Gray, A. J. (2010). Enabling ontology-based access to streaming data sources. In *The Semantic Web–ISWC 2010* (pp. 96–111). Springer.
Cao, J., Zhang, W., & Tan, W. (2012). Dynamic control of data streaming and processing in a virtualized environment. *IEEE Transactions on Automation Science and Engineering, 9*(2), 365–376.
Cochran, W. G. (2007). *Sampling techniques*. Wiley.
Komazec, S., Cerri, D., & Fensel, D. (2012). Sparkwave: continuous schema-enhanced pattern matching over rdf data streams. In *Proceedings of the 6th ACM International Conference on Distributed Event-Based Systems* (pp. 58–68).
Le-Phuoc, D., Dao-Tran, M., Parreira, J. X., & Hauswirth, M. (2011). A native and adaptive approach for unified processing of linked streams and linked data. In *The Semantic Web–ISWC 2011* (pp. 370–388). Springer.
Rete, C. (1982). A fast algorithm for the many pattern/many object pattern matching problem. *Artificial Intelligence, 19*, 17–37.
Vijayakumar, S., Zhu, Q., & Agrawal, G. (2010). Dynamic resource provisioning for data streaming applications in a cloud environment. In *Cloud Computing Technology and Science (CloudCom), 2010 IEEE Second International Conference on* (pp. 441–448). IEEE.
Vitter, J. S. (1985). Random sampling with a reservoir. *ACM Transactions on Mathematical Software (TOMS), 11*(1), 37–57.

Author Biographies

Yousra Chabchoub is currently an associate professor at the computer science department of the ISEP. She received, in 2005, her engineering degree from Télécom SudParis, and her Ph.D. degree from Pierre et Marie Curie University in 2009. Before joining ISEP, she was a postodoctoral researcher in Télécom-ParisTech. Her research interests include data mining, data stream management and semantic technologies.

Zakia Kazi-Aoul is an associate professor at ISEP since January 2009, member of the RDI (Research and Development in Information Technology) team and head of the Information Systems track. She began her career in research by integrating the Distributed Computing Master of Research at the University of Paris XI. After her Ph.D. at Telecom ParisTech, she performed a post-doc at Telecom SudParis in the MARGE team. Her research activities concern semantic large-scale data stream management.

Aliou Boly is currently Associate Professor at the University Cheikh Anta Diop de Dakar (UCAD), Senegal where he is responsible for the Master in business intelligence. He holds a Ph.D. in Computer Science from Telecom ParisTech obtained after a Master degree in computer science at the University of Paris Dauphine. His research interests include data warehousing, databases, data mining and data stream management.

Amadou Fall Dia is currently a Ph.D. student in RDI team at ISEP. He works directly in FUI Waves (waves-rsp.org), a project about semantic data stream management system where his current research focuses on scalable and distributed platform for processing semantic data streams. He holds a master degree in Business Intelligence from the University of Dakar (Senegal). His main areas of research include semantic web, data steam management, distributed systems, data mining, complex event processing, machine learning and data warehousing.

Efficiency Analysis of ASP Encodings for Sequential Pattern Mining Tasks

Thomas Guyet, Yves Moinard, René Quiniou and Torsten Schaub

Abstract This article presents the use of Answer Set Programming (ASP) to mine sequential patterns. ASP is a high-level declarative logic programming paradigm for high level encoding combinatorial and optimization problem solving as well as knowledge representation and reasoning. Thus, ASP is a good candidate for implementing pattern mining with background knowledge, which has been a data mining issue for a long time. We propose encodings of the classical sequential pattern mining tasks within two representations of embeddings (*fill-gaps* versus *skip-gaps*) and for various kinds of patterns: frequent, constrained and condensed. We compare the computational performance of these encodings with each other to get a good insight into the efficiency of ASP encodings. The results show that the *fill-gaps* strategy is better on real problems due to lower memory consumption. Finally, compared to a constraint programming approach (CPSM), another declarative programming paradigm, our proposal showed comparable performance.

1 Introduction

Pattern mining is a data analysis task aiming at identifying "meaningful" patterns in a database of structured data (e.g. itemsets, sequences, graphs). Sequential pattern mining consists in discovering subsequences as patterns in a sequence database (Shen et al. 2014). This problem has been introduced at the early beginning of the pattern

T. Guyet (✉)
AGROCAMPUS-OUEST/IRISA-UMR 6074, Rennes, France
e-mail: thomas.guyet@agrocampus-ouest.fr

Y. Moinard · R. Quiniou
Inria – Centre de Rennes, Rennes, France
e-mail: yves.moinard@inria.fr

R. Quiniou
e-mail: rene.quiniou@inria.fr

T. Schaub
Potsdam University, Potsdam, Germany
e-mail: torsten@cs.uni-potsdam.de

© Springer International Publishing AG 2018   41
B. Pinaud et al. (eds.), *Advances in Knowledge Discovery and Management*,
Studies in Computational Intelligence 732, https://doi.org/10.1007/978-3-319-65406-5_3

mining field (Agrawal and Srikant 1995) with the itemsets mining task (Agrawal et al. 1993). Sequential pattern mining is known to have a higher complexity than itemsets mining, but it has broad applications (Gupta and Han 2013). It includes—but is not limited to—the analysis of patient care pathways, education traces, digital logs (web access for client profiling, intrusion detection from network logs), customer purchase (rules for purchases recommendations), text and bioinformatic sequences.

In most cases, "interesting" patterns are the frequent ones. A pattern is said to be frequent if it appears at least f_{min} times in the database, where f_{min} is a frequency threshold given by the data analyst. This interestingness criterion reveals some important behaviours in the datasets and, above all, it benefits from an interesting property (anti-monotonicity) that make algorithms very efficient, even on large databases. Two decades of research on the specific task of frequent sequential pattern mining have led to very efficient mining methods and strategies to extract the complete set of frequent patterns or condensed representation of frequent patterns (Wang and Han 2004). These methods can currently process millions of sequences with very low frequency threshold.

The challenge of mining a deluge of data is about to be solved, but is also about to be replaced by another issue: the deluge of patterns. In fact, the size of the complete set of frequent patterns explodes with the database size and density (Lhote 2010). The data analyst cannot handle such volumes of results. A broad range of research, from data visualization (Perer and Wang 2014) to database sampling (Low-Kam et al. 2013) is currently attempting to tackle this issue. The data-mining community has mainly focused on the addition of expert constraints on sequential patterns (Pei et al. 2004).

Recent approaches have renewed the field of Inductive Logic Programming (Muggleton and De Raedt 1994) by exploring declarative pattern mining. Similarly, some works have tackled the itemset mining task (Guns et al. 2015; Järvisalo 2011). Recently, some propositions have extended the approach to sequence mining (Negrevergne and Guns 2015; Coquery et al. 2012; Métivier et al. 2013). Their practical use depends on the efficiency of their encoding to process real datasets. Thanks the improvements on satisfiability (SAT) or constraints programming (CP) solving techniques and solvers, such approaches become realistic alternatives for highly constrained mining tasks. Their computational performances closely reach those of dedicated algorithms.

The long term objective is to benefit from the genericity of solvers to let a user specify a potentially infinite range of constraints on the patterns. Thus, we expect to go from specific algorithm constraints to a rich query language for pattern mining.

The approach we propose in this paper uses the formalism of Answer Set Programming (ASP) and the solver *clingo*. ASP is a logic programming language, as Prolog. Its first order syntax makes ASP programs easy to understand. Furthermore, ASP benefits from efficient solvers to compute efficiently the solution answer sets (Lifschitz 2008).

The contributions of this article are twofold. (1) The article presents a declarative approach which provides a high-level specification of a broad range of sequential pattern mining tasks in a unique framework. It demonstrates that this mining task and its related problems—mining closed, maximal and constrain patterns—can easily be encoded with pure declarative ASP. (2) The article extensively evaluates the proposed encodings to draw the computational strengths and limits of ASP for declarative pattern mining. It gives also experimental results about time/memory computing efficiency of the solving process and provides alternative encodings to improve the computing efficiency. The proposed encodings are compared to the results of the CPSM software, based on CP programming (Negrevergne and Guns 2015).

The article is organized as follows: Sect. 2 introduces ASP programming, its principles and the solver *clingo*. Then in Sect. 3, we introduce sequential pattern mining. In Sect. 4, we give several ASP encodings of the basic sequential pattern mining task. Section 5 presents encodings for alternative sequential pattern mining tasks, including the use of constraints and the extraction of condensed representations. After presenting some related works in Sect. 6, we present our experiments in Sect. 7.

2 ASP—Answer Set Programming

In this section we introduce the Answer Set Programming (ASP) paradigm, syntax and tools. Section 2.1 introduces the main principles and notations of ASP. Section 2.2 illustrates them on the well-known graph coloring problem.

2.1 Principles of Answer Set Programming

ASP is a declarative programming paradigm. From a general point of view, declarative programming gives a description of what is a problem instead of specifying how to solve it. Several declarative paradigms have been proposed, differing in the modelling formalism they use. For instance, logic programming (Lallouet et al. 2013) specifies the problem using a logic formalism, the SAT paradigm encodes the problem with boolean expressions (Biere et al. 2009), the CP (constraint programming) paradigm specifies the problem using constraints (Rossi et al. 2006). ASP belongs to the class of logic programming paradigms, such as Prolog. The high-level syntax of logic formalisms makes generally the program easier to understand than other declarative programming paradigms.

An *ASP program* is a set of rules of the form

$$a_0 :\text{-} a_1, \ldots, a_m, \textbf{not } a_{m+1}, \ldots, \textbf{not } a_n. \tag{1}$$

where each a_i is a propositional atom for $0 \le i \le n$ and not stands for *default negation*. In the body of the rule, commas denote conjunctions between atoms. Contrary to Prolog, the order of atoms is meaningless. In ASP, rule (1) may be interpreted as "*if* $a_1, \ldots, a_m$ *are all true and if none of* $a_{n+1}, \ldots, a_n$ *can be proved to be true, then* a_0 *is true.*"

If $n = 0$, i.e. the rule body is empty, (1) is called a *fact* and the symbol "$:-$" may be omitted. Such a rule states that the atom a_0 has to be true. If a_0 is omitted, i.e. the rule head is empty, (1) represents an integrity constraint.

Semantically, a logic program induces a collection of so-called *answer sets*, which are distinguished models of the program determined by answer sets semantics; see Gelfond and Lifschitz (1991) for details. For short, a model assigns a truth value to each propositional atoms of the program and this set of assignments is valid. An answer set is a minimal set of true propositional atoms that satisfies all the program rules. Answer sets are said to be minimal in the way that only atoms that have to be true are actually true.

To facilitate the use of ASP in practice, several extensions have been developed. First of all, rules with *variables* are viewed as shorthands for the set of their ground instances. This allows for writing logic programs using a first order syntax. Such kind of syntax makes program shorter, but it hides the grounding step and its specific encoding issues, especially from the memory management point of view.

Further language constructs include *conditional literals* and *cardinality constraints* (Simons et al. 2002). The former are of the form

$$a : b_1, \ldots, b_m$$

the latter can be written as

$$s \{c_1; \ldots; c_n\} t$$

where a and b_i are possibly default negated literals for $0 \leq i \leq m$, and each c_j is a conditional literal for $1 \leq i \leq n$. The purpose of conditional literals is to govern the instantiation of a literal a through the literals $b_1, \ldots, b_m$. In a cardinality constraint, s (resp. t) provides the lower (resp. upper) bound on the number of literals from $c_1; \ldots; c_n$ that must be satisfied in the model.

A cardinality constraint in the head of the rule defines a specific rule called a *choice rule*:

$$s \{c_1; \ldots; c_n\} t :- a.$$

If a is true then all atoms of a subset $\mathscr{S} \subset \{c_1, \ldots, c_n\}$ of size between s and t have to be true. All such subsets are admissible according to this unique rule, but not in the same model. All such subsets contribute to alternative answer sets. It should be noted that alternative models are solved independently. It is not possible to specify constraints that involve several models.

ASP problem solving is ensured by efficient solvers (Lifschitz 2008) which are based on the same technologies as constraint programming solvers or satisfiability checking (SAT) solvers. *smodels* (Syrjänen and Niemelä 2001), *dlv* (Leone et al. 2006), *ASPeRiX* (Lefèvre and Nicolas 2009) or *clingo* (Gebser et al. 2011) are well-known ASP solvers. Due to the computational efficiency it has demonstrated and its broad application to real problems, we use *clingo* as a basic tool for designing our encodings.

Fig. 1 An example graph for
the graph coloring problem

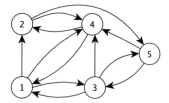

The basic method for programming in ASP is to follow a *generate-and-test* methodology. Choice rules generate solution candidates, while integrity constraints are tested to eliminate those candidates that violate the constraints. The programmer should not have any concern about how solutions are generated. He/she just has to know that all possible solutions will be actually evaluated. From this point of view, the ASP programming principle is closer to CP programming than to Prolog programming. Similarly to these declarative programming approaches, the difficulty of programming in ASP lies in the choices for the best way to encode problem constraints: it must be seen as the definition of the search space (*generate* part) or as an additional constraint on solutions within this search space (*test* part). This choices may have a large impact on the efficiency of the problem encoding.

2.2 A Simple Example of ASP Program

The following example illustrates the ASP syntax on encoding the graph coloring problem. Lines 9–10 specify the problem as general rules that solutions must satisfy while lines 1–6 give the input data that defines the problem instance related to the graph in Fig. 1.

The problem instance is a set of colors, encoded with predicates col/1 and a graph, encoded with predicates vertex/1 and edge/2. The input graph has 5 vertices numbered from 1 to 5 and 12 edges. The 5 fact-rules describing the vertex are listed in the same line (Line 2). It should be noted that, generally speaking, edge(1,2) is different from edge(2,1), but considering that integrity constraints for the graph coloring problem are symmetric, it is sufficient to encode directed edge in only one direction. Line 6 encodes the 3 colors that can be used: r, g and b. Lower case letters represent values, internally encoded as integers, while strings beginning with upper case letters represent variables (see line 9 for instance).

Lines 9 and 10 specify the graph coloring problem. The predicate color/2 encodes the color of a vertex: color(X,C) expresses that vertex X has color C. Line 10 is an integrity constraint. It forbids neighbor vertices X and Y to have the same color C.[1] The ease of expressing such integrity constraints is a major feature of ASP.

---

[1] It is important to notice that the scope of a variable is the rule and each occurrence of a variable in a rule represents the same value.

```
1 % instance of the problem: the graph and colors
2 vertex(1). vertex(2). vertex(3). vertex(4). vertex(5)
    .
3 edge(1,2). edge(1,3). edge(1,4). edge(2,4). edge(2,5)
    .
4 edge(3,1). edge(3,4). edge(3,5). edge(4,1). edge(4,2)
    .
5 edge(5,3). edge(5,4).
6 col(r). col(b). col(g).
7
8 % graph coloring problem specification
9 1 { color(X, C) : col(C) } 1 :- vertex(X).
10 :- edge(X, Y), color(X, C), color(Y, C).
```
Listing 3.1 Encoding of graph coloring—ASP syntax and encoding example

Line 9 is a choice rule indicating that for a given vertex X, an answer set must contain exactly one atom of the form `color(X,C)` where C is a color. The grounded version of this rule is the following:

```
1 { color(1, r), color(1, b), color(1, g) } 1.
1 { color(2, r), color(2, b), color(2, g) } 1.
1 { color(3, r), color(3, b), color(3, g) } 1.
1 { color(4, r), color(4, b), color(4, g) } 1.
1 { color(5, r), color(5, b), color(5, g) } 1.
```

The variable X is expanded according to the facts in line 2 and for each vertex, a specific choice rule is defined. Within the brackets, the variable C is expanded according to the conditional expression in the rule head of line 9: the only admissible values for C are color values. For each line of the grounded version of the program, one and only one atom within brackets can be chosen. This corresponds to a unique mapping of a color to a vertex. Line 9 can be seen as a search space generator for the graph coloring problem.

The set `color(1,b) color(2,r) color(3,r) color(4,g) color(5,b)` is an answer set for the above program (among several others).

For more detailed presentation of ASP programming paradigm, we refer the reader to recent article of Janhunen and Nimeläthe (2016).

2.3 The Potassco collection of ASP tools

The *Potassco* collection is a set of tools for ASP developed at the University of Potsdam. The main tool of the collection is the ASP solver *clingo* (Gebser et al. 2011). This solver offers both a rich syntax to facilitate encodings[2] and a good solving efficiency. It is worth-noting that the ASP system *clingo* introduced many

---

[2]*clingo* is fully compliant with the recent ASP standard: https://www.mat.unical.it/aspcomp2013/ASPStandardization.

facilities to accelerate the encoding of ASP programs. For the sake of simplicity, we do not use them in the presented programs. A complete description of the *clingo* syntax can be found in Gebser et al. (2014).

The *clingo* solving process follows two consecutive main steps:

1. *grounding* transforms the initial ASP program into a set of propositional clauses, cardinality constraints and optimisation clauses. Note that grounding is not simply a systematic problem transformation. It also simplifies the rules to generate the as short as possible equivalent grounded program.
2. *solving* consists in finding from one to all solutions of the grounded program. This step is performed by *clasp* which is a conflict-driven ASP solver. The primary *clasp* algorithm relies on conflict-driven *nogood* learning. It is further optimized using sophisticated reasoning and implementation techniques, some specific to ASP, others borrowed from CDCL-based SAT solvers.

The overall process may be controlled using procedural languages, e.g. *Python* or *lua* (Gebser et al. 2014). These facilities are very useful to automate processes and to collect statistics on solved problems. Despite this procedural control which enables to interact with the grounder or the solver, it is important to note that once a program has been grounded, it cannot be changed.

3 Sequential Pattern Mining: Definition and Notations

Briefly, the sequential pattern mining problem consists in retrieving from a sequence database $\mathscr{D}$ every frequent non empty sequence P, so-called a sequential pattern. Sequences, either in $\mathscr{D}$ or sequential patterns P, are multiset sequences of itemsets over a fixed alphabet of symbols (also called items). A pattern is frequent if it is a subsequence of at least f_{min} sequences of $\mathscr{D}$, where f_{min} is an arbitrary given threshold. In this section, we introduce classical definitions and notations for frequent sequential pattern mining which will be useful to formulate the problem in an ASP setting. In the sequel, if not specified otherwise, a pattern is a sequential pattern.

3.1 Sequences

We introduce here the basic definitions of sequences of itemsets. $[n] = \{1, \ldots, n\}$ denotes the set of the first n strictly positive integers.

Let $(\mathscr{I}, =, <)$ be the set of items (alphabet) with a total order (e.g. lexicographic order). An *itemset* $A = (a_1, a_2, \ldots, a_n)$ is a subset of distinct increasingly ordered items from $\mathscr{I}$: $\forall i \in [n]$, $a_i \in \mathscr{E} \wedge \forall i \in [n-1]$, $a_i < a_{i+1} \wedge \forall i, j \in [n]$, $i \neq j \implies a_i \neq a_j$. An itemset $\beta = (b_i)_{i \in [m]}$ is a sub-itemset of $\alpha = (a_i)_{i \in [n]}$, denoted $\beta \sqsubseteq \alpha$, iff there exists a sequence of m integers $1 \leq i_1 < i_2 < \cdots < i_m \leq n$ such that $\forall k \in [m], b_k = a_{i_k}$. A *sequence* S is an ordered set of itemsets $S = \langle s_1, s_2, \ldots, s_n \rangle$:

$\forall i, j \in [n]$, $i < j$ means that s_i occurs before s_j. The length of sequence S, denoted $|S|$, is equal to its number of itemsets. Two sequences $S = \langle s_1, s_2, \dots, s_n \rangle$ and $T = \langle t_1, t_2, \dots, t_m \rangle$ are equal iff $n = m$ and $\forall i \in [n]$, $s_i = t_i$.

$T = \langle t_1, t_2, \dots, t_m \rangle$ is a *sub-sequence* of $S = \langle s_1, s_2, \dots, s_n \rangle$, denoted $T \preceq S$, iff there exists a sequence of integers $1 \leq i_1 < i_2 < \cdots < i_m \leq n$ such that $\forall k \in [m]$, $t_k \sqsubseteq s_{i_k}$. In other words, $(i_k)_{1 \leq k \leq m}$ defines a mapping from $[m]$, the set of indexes of T, to $[n]$, the set of indexes of S. We denote by $T \prec S$ the strict sub-sequence relation such that $T \preceq S$ and $T \neq S$.

$T = \langle t_1, t_2, \dots, t_m \rangle$ is a *prefix* of $S = \langle s_1, s_2, \dots, s_n \rangle$, denoted $T \preceq_b S$, iff $\forall i \in [k-1]$, $t_i = s_i$ and $t_m \sqsubseteq s_m$. Thus, we have $T \preceq_b S \implies T \preceq S$.

A sequence T *is supported* by a sequence S if T is a sub-sequence of S, i.e. $T \preceq S$.

Example 1 (Sequences, subsequences and prefixes) Let $\mathscr{I} = \{a, b, c\}$ with a lexicographic order ($a < b$, $b < c$) and the sequence $S = \langle a(bc)(abc)cb \rangle$. To simplify the notation, parentheses are omitted around itemsets containing only one item. The length of S is 5. $\langle abb \rangle$, $\langle (bc)(ac) \rangle$ or $\langle a(bc)(abc)cb \rangle$ are sub-sequences of S. a, $\langle a(bc) \rangle$ and $\langle a(bc)a \rangle$ are prefixes of S.

Proposition 1 $\prec$ *and* $\prec_b$ *induces two partial orders on the sequence set. For all sequences* (s, s'), $s \prec_b s' \implies s \prec s'$.

3.2 Sequential Pattern Mining

Let $\mathscr{D} = \{S_1, S_2, \dots, S_N\}$ be a set of sequences. $\mathscr{D}$ is called a *sequence database*. The *support of a sequence* S in $\mathscr{D}$, denoted by $supp_{\mathscr{D}}(S)$, is the number of sequences of $\mathscr{D}$ that support S:

$$supp_{\mathscr{D}}(S) = |\{S_i \in \mathscr{D} | S \prec S_i\}|$$

Proposition 2 $supp_{\mathscr{D}}(.)$ *is an anti-monotonic measure on the set of subsequences of a sequence database* $\mathscr{D}$ *structured by* $\prec$ *or* $\prec_b$.

This proposition implies that for all pairs of sequences P, Q:

$$P \prec Q \implies supp_{\mathscr{D}}(P) \geq supp_{\mathscr{D}}(Q), \text{ and}$$

$$P \prec_b Q \implies supp_{\mathscr{D}}(P) \geq supp_{\mathscr{D}}(Q).$$

Let f_{min} be a *frequency threshold* defined by the analyst. For any sequence S, if $supp_{\mathscr{D}}(S) \geq f_{min}$, we say that S is a *frequent sequence* or a *(frequent) sequential pattern* of $\mathscr{D}$. *Mining sequential patterns* consists in extracting all frequent subsequences in a sequence database $\mathscr{D}$.

Every pattern mining algorithm (Agrawal and Srikant 1995; Wang and Han 2004; Pei et al. 2007) uses the anti-monotonicity property to browse efficiently the pattern search space. In fact, this property ensures that a sequence P including a sequence

Q which is not frequent, cannot be frequent itself. So, the main idea of classical algorithms is to extend the patterns until identifying a non frequent pattern.

Example 2 (Sequential pattern mining) To illustrate the concepts introduced above, we consider the following sequence database $\mathcal{D}$ containing sequences built on items in $\mathcal{I} = \{a, b, c, d\}$ such that $a < b$, $b < c$ and $a < c$. In this running example, and in the rest of this article, we focus on *simple* sequences of items instead of sequences of itemsets.

SeqId	Sequence
1	$\langle ac \rangle$
2	$\langle dabc \rangle$
3	$\langle b \rangle$
4	$\langle abc \rangle$
5	$\langle ab \rangle$
6	$\langle acbc \rangle$
7	$\langle abc \rangle$

Given a threshold value $f_{min} = 3$, the frequent sequential patterns are: $\langle a \rangle$, $\langle b \rangle$, $\langle c \rangle$, $\langle ac \rangle$, $\langle bc \rangle$, $\langle ab \rangle$ and $\langle abc \rangle$.

It is interesting to relate the sequential pattern mining task with the presentation of ASP principles. The sequential pattern mining task rises two issues: (1) exploring a large search space, i.e. the potentially infinite set of sequences and (2) assessing the frequency constraint (with respect to the given database). Thus, sequential pattern mining can be considered as a *generate and test* process which makes it straightforward to encode the mining task using ASP principles: (1) choice rules will define the search space and (2) the frequency assessment will be encoded using integrity constraints.

4   Frequent Sequential Pattern Mining with ASP

In this section, we present several ASP encodings for sequential pattern mining. We assume that the database contains sequences of itemsets. But for the sake of simplicity, we will restrict patterns to sequences of items (each itemset is a singleton). Listing 3.10 in Appendices gives an encoding for the full general case of sequential pattern mining.

Our proposal is borrowed from Järvisalo's (2011): the solution of the sequential pattern mining ASP program is all the answer sets (AS), each of which contains the atoms describing a single frequent pattern as well as its occurrences in database sequences. The solution relies on the "generate and test principle": generate combinatorially all the possible patterns and their related occurrences in the database sequences and test whether they satisfy the specified constraints.

4.1 Modelling Database, Patterns and Parameters

A sequence database $\mathscr{D}$ is modelled by the predicate `seq(T,Is,I)` which holds if sequence `T` contains item `I` at index `Is`.

Example 3 (A sequence database encoded in ASP) The following facts encode the database of Example 2:

```
seq(1,1,a).  seq(1,2,c).
seq(2,1,d).  seq(2,2,a).  seq(2,3,b).  seq(2,4,c).
seq(3,1,b).
seq(4,1,a).  seq(4,2,b).  seq(4,3,c).
seq(5,1,a).  seq(5,2,b).
seq(6,1,a).  seq(6,2,c).  seq(6,3,b).  seq(6,4,c).
seq(7,1,a).  seq(7,2,b).  seq(7,3,c).
```

Similarly, the current pattern is modelled by the predicate `pat(Ip,I)` which holds if the current pattern contains item `I` at index `Ip`.

For example, the pattern $\langle abc \rangle$ is modelled by the following atoms:

`pat(1,a). pat(2,b). pat(3,c).`

In addition, we define two program constants:

- `#const th=23.` represents f_{min}, the minimal frequency threshold, i.e. the requested minimal number of supporting sequences
- `#const maxlen=10.` represents the maximal pattern length.

Let S be a set of ground atoms and $P \subset S$ the set of `pat(Ip,I)` atoms in S, according to the Järvisalo's encoding principle we would like an ASP program π such that S is an answer set of π iff the pattern defined by P is a frequent sequential pattern in the database $\mathscr{D}$.

4.2 Two Encodings for Sequential Pattern Mining

The main difficulty in declarative sequential pattern mining is to decide whether a pattern $P = \langle p_1, \ldots, p_n \rangle$ supports a sequence $S = \langle s_1, \ldots, s_m \rangle$ of the database. According to Definition 1, it means that it exists a mapping $e = (e_i)_{1 \leq i \leq n}$ such that $p_i = s_{e_i}$. Unfortunately, this definition is not usable in practice to implement an efficient ASP encodings. The difficulty comes from the possible multiple mappings of a pattern in a single sequence. On the other hand, the detailed mapping description is not required here, we simply have to defined embeddings that exists iff a pattern supports a sequence. An embedding of a pattern in a sequence is given by the description of a relation between pattern item indexes to sequence item indexes.

This section presents two encodings of sequential pattern mining. These two encodings differ in their representation of embeddings, as illustrated in Fig. 2. Two embedding strategies have been defined and compared in our results: *skip-gaps* and *fill-gaps*.

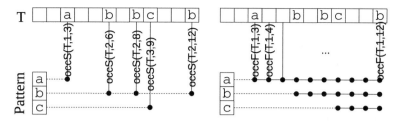

Fig. 2 Illustration of embeddings strategies. On the *left*, *skip-gaps*, on the *right*, *fill-gaps*

More formally, let $P = \langle p_1, \ldots, p_n \rangle$ be a pattern sequence and $S_T = \langle s_1, \ldots, s_m \rangle$ be the T-th sequence of $\mathcal{D}$. In the *skip-gaps* strategy, an embedding $\mathcal{E}$ is a relation over $[1, m] \times [1, n]$ such that $\forall (i, j) \in \mathcal{E}$, $i \leq j \wedge p_i = s_j$ and $\forall (i, j), (i', j') \in \mathcal{E}$, $i < i' \implies j < j'$. In the *fill-gaps* strategy, an embedding $\mathcal{E}'$ is the same relation as $\mathcal{E}$ (i.e. $(i, j) \in \mathcal{E} \implies (i, j) \in \mathcal{E}'$) with the additional specification: $\forall i \in [1, m]$, $\forall j, j' \in [1, n]$, $(i, j) \in \mathcal{E}' \wedge j < j' \implies (i, j') \in \mathcal{E}'$. This additional specification expresses that once a pattern item has been mapped to the leftmost (having the lowest index, let it be j), the knowledge of this mapping is maintained on remaining sequences items with indexes $j' > j$. So, a *fill-gaps* embedding makes only explicit the "leftmost admissible matches" of P items in sequence S_T.

Relations $\mathcal{E}$ and $\mathcal{E}'$ are interesting because (i) that can be computed in a constructive way (i.e. without encoding guesses) and (ii) they contains the information required to decide whether the pattern supports the sequence.

The two following sections detail the ASP programs for extracting patterns under each embedding strategy.

4.2.1 The Skip-Gaps Approach

In the first ASP encoding, an embedding of the current pattern $P = \langle p_i \rangle_{1 \leq i \leq n}$ in sequence $T = \langle s_i \rangle_{1 \leq i \leq m}$ is described by a set of atoms occS(T,Ip,Is) which holds if Ip-th the pattern item (occurring at index Ip) is identical to the Is-th item in sequence T (formally, $p_{\text{Ip}} = s_{\text{Is}} \wedge \langle p_i \rangle_{1 \leq i \leq \text{Ip}} \prec \langle s_i \rangle_{1 \leq i \leq \text{Is}}$). The set of valid atoms occS(T,_,_) encodes the relation $\mathcal{E}$ above and is illustrated in Fig. 2 (on the left).

Example 4 (Illustration of skip-gaps embedding approach) Let $P = \langle ac \rangle$ be a pattern represented by pat(1,a).pat(2,c). Here follows, the embeddings of pattern P in the sequences of Example 2:

```
occS(1,1,1)  occS(1,2,2)
occS(2,1,2)  occS(2,2,4)
occS(4,1,1)  occS(4,2,3)
occS(5,1,1)
occS(6,1,1)  occS(6,2,2)  occS(6,2,4)
occS(7,1,1)  occS(7,2,3)
```

```
1 item(I) :- seq(_, _,I).
2
3 %sequential pattern generation
4 patpos(1).
5 0 { patpos(Ip+1) } 1 :- patpos(Ip), Ip<maxlen.
6 patlen(L) :- patpos(L), not patpos(L+1).
7
8 1 { pat(Ip,I): item(I) } 1 :- patpos(Ip).
9
10 %pattern embeddings
11 occS(T,1    ,Is):- seq(T,Is,I), pat(1,    I).
12 occS(T,Ip+1,Is):- seq(T,Is,I), pat(Ip+1,I), occS(T,Ip,Js), Js<Is.
13
14 %frequency constraint
15 support(T) :- occS(T,L,_), patlen(L).
16 :- { support(T) } < th.
```

Listing 3.2 Encoding of frequent sequential pattern mining—skip-gaps strategy

The pattern could not be fully identified in the fifth sequence. There are two possible embeddings in the sixth sequence. Atom occS(6,1,1) is used for both. Nonetheless, this sequence must be counted only once in the support.

Listing 3.2 gives the ASP program for sequential pattern mining. The first line of the program is called a *projection*. It defines a new predicate that provides all items from the database. The symbol "_" denotes an anonymous (or don't care) variable.

Lines 4–8 of the program encode the pattern generation. Predicate patpos/1 defines the allowed sequential pattern indexes, beginning at index 1 (line 4). Line 5 is a choice rule that generates the successive pattern positions up to an ending index iterating from 2 to maxlen: patpos(Ip+1) is true if there is a pattern position at index Ip and Ip is lower than maxlen. Line 6 defines the length of a pattern: patlen(L) holds if L is the index of the last pattern item (there is no pattern item with a greater index). This predicate is used to decide whether an embedding has been completed or not. Finally, line 8 is a choice rule that associates exactly one item with each position x. We can note that each possible sequence is generated once and only once. So, there is no redundancy in the search space exploration.

Lines 11–12 encode pattern embedding search. Line 11 guesses a sequence index for the first pattern item: occS(T,1,Is) holds if the first pattern item is identical to the Is-th of sequence T (i.e. $p_1 = s_{Is}$). Line 12 guesses sequence indexes for pattern items at indexes strictly greater than 1. occS(T,Ip,Is) holds if the Ip-th pattern item is equal to the Is-th sequence item (i.e. $p_{Ip} = s_{Is}$) and the preceding pattern item is mapped to a sequence item at an index strictly lower than Is. Formally, this rule expresses the following implication (Jp, Is -1) $\in \mathscr{E} \wedge p_{Ip} = s_{Is} \wedge$ Ip $>$ Jp $\implies$ (Ip, Is) $\in \mathscr{E}$ and recursively, we have $\langle p_i \rangle_{1 \leq i \leq Ip} \prec \langle s_i \rangle_{1 \leq i \leq Is}$. It should be noted that this encoding generates all the possible embeddings of some pattern.

Finally, lines 15–16 are dedicated to assess the pattern frequency constraint. support(T) holds if the database sequence T supports the pattern, i.e. if an atom occS holds for the last pattern position. The last line of the program is an integrity constraint ensuring that the number of supported sequences is not lower than the

```
10 %pattern embeddings
11 occF (T,1,Is)    :- seq (T,Is,I), pat (1,I).
12 occF (T,Ip,Is)   :- occF (T, Ip-1, Is-1), seq (T,Is,I), pat (L,I).
13 occF (T,Ip,Is)   :- occF (T, Ip, Is-1), seq (T,Is,_).
14
15 %frequency constraint
16 seqlen (T,L)   :- seq (T,L,_), not seq (T,L+1,_).
17 support (T)    :- occF (T, L, LS), patlen (L), seqlen (T,LS).
18 :- { support (T) } < th.
```

Listing 3.3 Encoding of frequent sequential pattern mining—fill-gaps strategy (see Listing 3.2 for first lines).

threshold `th` or, in other words, that the support of the current pattern is greater than or equal to the threshold.

4.2.2 The Fill-Gaps Approach

In the *fill-gap* approach, an embedding of the current pattern P is described by a set of atoms `occF(T,Ip,Is)` having a slightly different semantics than in the *skip-gap* approach. `occF(T,Ip,Is)` holds if at sequence index `Is` it is true that the `Ip`-th pattern item has been mapped (to some sequence index equal to `Is` or lower than `Is` if `occF(T,Ip,Is-1)` holds). More formally, we have $\langle p_i \rangle_{1 \le i \le Ip} \prec \langle s_i \rangle_{1 \le i \le Is}$. The set of atoms `occF(T,_,_)` encodes the relation $\mathscr{E}'$ above and is illustrated in Fig. 2 (on the right).

Example 5 (Fill-gaps approach embedding example) Pattern $P = \langle a, c \rangle$ has the following fill-gaps embeddings (represented by occF atoms) in the sequences of the database of Example 2:

```
occF (1,1,1)  occF (1,1,2)  occF (1,2,2)
occF (2,1,2)  occF (2,1,3)  occF (2,1,4)  occF (2,2,4)
occF (4,1,1)  occF (4,1,2)  occF (4,1,3)  occF (4,2,3)
occF (5,1,1)  occF (5,1,2)
occF (6,1,1)  occF (6,1,2)  occF (6,1,3)  occF (6,1,4)  occF (6,2,2)  occF
      (6,2,3)  occF (6,2,4)
occF (7,1,1)  occF (7,1,2)  occF (7,1,3)  occF (7,2,3)
```

Contrary to the skip-gap approach example (see Example 4), the set of `occF(T,Ip,Is)` atoms alone is not sufficient to deduce all occurrences. For instance, occurrence with indexes $(3, 8, 9)$ is masked.

Listing 3.3 gives the ASP program for sequential pattern mining with the fill-gaps strategy. The rules are quite similar to those encoding the skip-gaps method. The main difference comes from the computation of embeddings (lines 11–13). As in Listing 3.2, line 11 guesses a sequence index for the first pattern item: `occF(T,1,Is)` holds if the first pattern item is identical to the `Is`-th of sequence `T` (i.e. $p_{Ip} = s_{Is}$).

Line 12 guesses sequence indexes for pattern items at indexes strictly greater than 1. `occS(T,Ip,Is)` holds if the `Ip`-th pattern item is equal to the `Is`-th sequence item and the preceding pattern item is mapped to some sequence item at some index strictly lower than $\mathtt{Is}$. More formally, we have that $p_{\mathtt{Ip}} = s_{\mathtt{Is}} \wedge (\mathtt{Ip} - 1, \mathtt{Is} - 1) \in \mathcal{E}' \implies (\mathtt{Ip}, \mathtt{Is}) \in \mathcal{E}'$.

Line 13 simply maintains the knowledge that the $\mathtt{Ip}$-th pattern item has been mapped all along the further sequence indexes, i.e. $\mathtt{occF}(\mathtt{T}, \mathtt{Ip}, \mathtt{Is})$ holds if $\mathtt{occF}(\mathtt{T}, \mathtt{Ip}, \mathtt{Is\text{-}1})$ holds. More formally, $(\mathtt{Ip} - 1, \mathtt{Is} - 1) \in \mathcal{E}' \implies (\mathtt{Ip}, \mathtt{Is}) \in \mathcal{E}'$. In combination with previous rules, we thus have recursively that $\mathtt{occF}(\mathtt{T}, \mathtt{Ip}, \mathtt{Is})$ is equivalent to $\langle p_i \rangle_{1 \leq i \leq \mathtt{Ip}} \prec \langle s_i \rangle_{1 \leq i \leq \mathtt{Is}}$.

Line 17 a sequence is supported by the pattern an $\mathtt{occF}$ atoms exists at the last position $\mathtt{LS}$ of the sequence, computed line 16. The remaining rules for testing whether it is greater than the threshold $\mathtt{th}$ are identical to those in the skip-gaps approach.

4.3 Sequential Pattern Mining Improvements

The main objective of this section is to present alternative encodings of the sequential pattern mining task. These encodings attempt to take advantage of known properties of the sequential pattern mining task to support the solver to mine datasets more efficiently or with less memory requirements. The efficiency of these improvements will be compared in the experimental part.

4.3.1 Filter Out Unfrequent Items

The first improvement consists in generating patterns from only frequent items. According to the anti-monotonicity property, all items in a pattern have to be frequent. The rules in Listing 3.4 may replace the projection rule previously defining the available items. Instead, an explicit aggregate argument is introduced to evaluate the frequency of each item $\mathtt{I}$ and to prune it if it is unfrequent.

In the new encoding, the predicate $\mathtt{sitem/1}$ defines the set of items that occurs in the database and $\mathtt{item/1}$ defines the frequent items that can generate patterns.

```
1 sitem(I)  :-  seq(_, _,I).
2 item(I)   :-  sitem(I), #count{ T:seq(T,_,I) } >= th.
```

Listing 3.4 Restriction of the pattern traversal to sequences made of frequent items only

4.3.2 Using Projected Databases

The idea of this alternative encoding is to use the principle of *projected databases* introduced by algorithm PrefixSpan (Pei et al. 2004). Let $P = \langle p_1, \ldots, p_n \rangle$ be a pattern, the projected database of $\mathcal{D} = \{S_1, \ldots, S_n\}$ is $\{S'_1, \ldots, S'_n\}$ where S'_i is the projected sequence of S_i with respect to P. Let $S_i = \langle s_1, \ldots, s_m \rangle$ be a sequence.

Then the projected sequence of S_i is $S'_i = \langle s_{k+1}, \dots s_m \rangle$ where k is the position of the last item of the first occurrence of P in S_i. If P does not occur in S_i then S'_i is empty.

A projected database is smaller than the whole database and the set of its frequent items is consequently much smaller than the original set of frequent items. The idea is to improve the candidate generation part of the algorithm by making use of items from projected databases. Instead of generating a candidate (a sequential pattern) by extending a frequent pattern with an item that is frequent in the whole database, the pattern extension operation uses only the items that are frequent in the database projected along this pattern.

```
8  item(1,I)       :- sitem(I),
9                     #count{ T: seq(T,_,I) } >= th.
10 item(Ip+1,I):- item(Ip,I),
11                    #count{ T: seq(T,Js,I),occS(T,Ip,Is),Js>Is }>= th
12 1 { pat(Ip,I)  :  item(Ip,I) } 1 :- patpos(Ip).
```
Listing 3.5 Pattern generation with prefix-projection principle—skip-gaps strategy

```
8  item(1,I)       :- sitem(I),
9                     #count{ T: seq(T,_,I) } >= th.
10 item(Ip+1,I)  :- item(Ip,I),
11                    #count{ T: seq(T,Is,I), occF(T,Ip,Is) } >= th.
12 1 { pat(Ip,I)  :  item(Ip,I) } 1 :- patpos(Ip).
```
Listing 3.6 Pattern generation with prefix-projection principle—fill-gaps strategy

The ASP encoding of the prefix-projection principle is given in Listing 3.5 for the skip-gaps strategy and in Listing 3.6 for the fill-gaps strategy. The programs of Listings 3.2 and 3.3 remain the same except for the generation of patterns defined by `patpos/1` and the new predicate `item/2`. `item(Ip,I)` defines an item `I` that is frequent in sequence suffixes remaining after removing the prefix of the sequence containing the first occurrence of the `x-1`-pattern prefix (consisting of the `Ip-1` first positions of the pattern). Lines 8–9 are similar to those in Listing 3.4. `item(1,I)` defines the frequent items, i.e. those that are admissible as first item of a frequent pattern. Lines 10–11 generates the admissible items for pattern position `Ip+1`. Such an item must be admissible for position `Ip` and be frequent in sequence suffixes (sub-sequence after at least one (prefix) pattern embedding). For skip-gaps, the sequence suffix is defined by `seq(T,Js,I)`, `occS(T,Ip,Is)`, `Js>Is` (the items at sequence positions farther away than the last position that matches the last (partial) pattern item at position `Ip`). For fill-gaps, `seq(T,Js,I)`, `occF(T,Ip,Is)` is sufficient because `occF(T,Ip,Is)` atoms represent the sequence suffix beginning at the sequence position that matches the last (partial) pattern item (at position `Ip`).

4.3.3 Mixing Itemsets and Sequences Mining

In Järvisalo (2011), Järvisalo showed that ASP can be efficient for itemset pattern mining. The main idea of this last alternative approach is to mine frequent itemsets and to derive sequential patterns from them.

```
 1 sitem(I) :- seq(_, _,I).
 2
 3 % extract frequent itemsets
 4 db(T,I) :- seq(T,_,I).
 5 0 { in_itemset(I) } 1 :- th { in_support(T):db(T,I) }, sitem(I).
 6 in_support(T) :- 0 { conflict_at(T,I) : sitem(I) } 0, db(T,_).
 7 conflict_at(T,I) :- not db(T,I), in_itemset(I), db(T,_).
 8
 9 %sequential pattern generation from frequent itemsets
10 patpos(1).
11 0 { patpos(Ip+1) } 1 :- patpos(Ip), Ip<maxlen.
12 patlen(L) :- patpos(L), not patpos(L+1).
13
14 1 { pat(Ip,I) : in_itemset(I) } 1 :- patpos(Ip).
15
16 :- 0 { pat(Ip,I) : patpos(Ip) } 0 , in_itemset(I).
17
18 %pattern occurrences
19 occS(T,1,Is)     :- in_support(T), seq(T,Is,I), pat(1,   I).
20 occS(T,Ip+1,Is):- occS(T,Ip,Js), seq(T,Is,I), pat(Ip+1,I), Js<Is
                 .
21
22 support(T) :- occS(T, L, _), patlen(L).
23 :- { support(T) } < th.
```

Listing 3.7 Mining frequent sequences from frequent itemsets—skip-gaps strategy

This time, the itemset mining step extracts a frequent itemset pattern $I = (e_i)_{i \in [n]}$, $e_i \in \mathcal{I}$. A sequential pattern $S = (s_i)_{i \in [m]}$ is generated using the items of the itemset, i.e. $\forall i \in [m], \exists j \in [n], s_i = e_j$ taking into account that items may be repeated within a sequential pattern and that every item from I must appear in S. If not, there would exist a subset $J \subset I$ that would generate the same sequence s. This would lead to numerous redundant answer sets for similar frequent sequences and would cause a performance drop.

Listing 3.7 gives the entire encoding of this alternative for the skip-gaps strategy.[3] Rules in Lines 4–7 extract frequent itemsets, represented by the predicate in_itemset/1, borrowed from Järvisalo's encoding (Järvisalo 2011). Next, the generation of sequential patterns in line 14 uses only items from such a frequent itemset. Line 16 defines a constraint required to avoid answer set redundancies. The remaining part of the program is left unchanged.

5   Alternative Sequential Pattern Mining Tasks

In this section, we illustrate how the previous encodings can be modified to solve more complex mining tasks. Our objective is to show the impressive expressiveness of ASP which let us encode a very wide range of mining tasks. We focus our attention on

---

[3]A similar encoding can be done for the fill-gaps strategy applying the same changes as above.

the most classical alternative sequential pattern mining tasks: constrained sequential patterns and condensed representation of sequential patterns.

In Negrevergne and Guns (2015), the authors organize the constraints on sequential patterns in three categories: (1) constraints on patterns, (2) constraints on patterns embeddings, (3) constraints on pattern sets. These constraints are provided by the user and capture his background knowledge.

The following section shows that our ASP approach enables to add constraints on individual patterns (constraints of categories 1 and 2). But, as ASP cannot compare models with each others, the third category of constraints can not be encoded directly.

In Sect. 5.2, we transform the classical definition of the most known constraints of the third category—the condensed representations—to encode them in pure ASP. Condensed representations (maximal and closed patterns) have been widely studied due to their monotonicity property, and to their representativeness with respect to frequent patterns. Concerning more generic constraints on pattern sets, such as the extraction of skypatterns (Ugarte et al. 2015), we have proposed in Gebser et al. (2016) an ASP-based approach for mining sequential skypatterns using *asprin* for expressing preferences on answer sets. *asprin* (Brewka et al. 2015) provides a generic framework for implementing a broad range of preferences relations on ASP models and can easily manage them. This approach is out of the scope of this article.

5.1 Constraints on Patterns and Embeddings

Pei et al. (2007) defined seven types of constraints on patterns and embeddings. In this section, we describe each of these constraints keeping their original numbering. Constraints 1, 2, 3 and 5 are pattern constraints, while constraints 4, 6 and 7 are embedding constraints. If not stated otherwise, the base encoding is the skip-gaps strategy and line numbers refers to Listing 3.2.

In a first approach, constraints on patterns and on embeddings may be trivially encoded by adding integrity constraints. But these integrity constraints acts a posteriori, during the test stage, for invalidating candidate models. A more efficient method consists in introducing constraints in the generate stage, specifically in choice rules, for pruning the search space early.

Constraint 1—Item constraint. An item constraint specifies what are the particular individual or groups of items that should or should not be present in the patterns. For instance, the constraint "patterns must contain item 1 but not item 2 nor item 3" can be encoded using `must_have/1` and `cannot_have/1` predicates: `must_have(1). cannot_have(2). cannot_have(3).`

To cope with this kind of constraint, Line 8 of Listing 3.2 is modified as:

```
8 1 { pat(X,I): item(I), not cannot_have(I) } 1 :- patpos(X).
9 :- { pat(X,I) : must_have(I) } < 1.
```

The encoding of Line 8 modifies the choice rule to avoid the generation of known invalid patterns, i.e. patterns with forbidden items. Line 9 is a new constraint that imposes to have at least one of the required items.

Constraint 2—Length constraint. A length constraint specifies a prerequisite on pattern length. The maximal length constraint is anti-monotonic while the minimal length is not anti-monotonic. The maximal length constraint is already encoded using the program constant `maxlen` in our encodings. A new constant `minlen` is defined to encode the minimal length constraint and a new rule is added to predicate `patpos/1` to impose at least `minlen` positions in patterns instead of only one.

```
#const minlen = 2.
patpos(1).
patpos(X+1) :- patpos(X), X<=minlen.
0 { patpos(X+1) } 1 :- patpos(X), X<maxlen.
```

Constraint 3—Super-pattern constraint. A super-pattern constraint enforces the extraction of patterns that contain one or more given sub-patterns. Mandatory sub-patterns are defined by means of the new predicate `subpat(SP,P,I)` expressing that sub-pattern `SP` contains item `I` at position `P`.

Predicate `issubpat(SP)` verifies that the sub-pattern `SP` is included in the pattern. An approach similar to embedding computation may be used:

```
issubpat(SP,1,P) :- pat(P,I), subpat(SP,1,I).
issubpat(SP,Pos+1,P) :- issubpat(SP,Pos,Q), pat(P,I),
                        subpat(SP,Pos+1,I), Q<P.
issubpat(SP) :- issubpat(SP,L,_), subpatlen(SP,L).
```

`issubpat(SP)` is true if the sub-pattern `SP` is a sub-pattern of the current pattern. This predicate is used to define the final integrity constraint:

```
:- #count{ SP : issubpat(SP), subpat(SP,_,_) } = 0.
```

Constraint 4—Aggregate constraint. An aggregate constraint is a constraint on an aggregation of items in a pattern, where the aggregate function can be *sum*, *avg*, *max*, *min*, *standard deviation*, etc. The only aggregates that are provided by *clingo* are #sum, #max and #min. For example, let us assume that to each item `I` is assigned a cost `C`, which is given by predicate `cost(I,C)`. The following constraint enforces the selection of patterns having a total cost of at least 1000.

```
:- #sum{ C,X : cost(I,C), pat(X,I) } < 1000.
```

As an integrity constraint, this rule means that it is not possible to have a total amount lower than 1000 for pattern. It should be noted that `C` values are summed for each pair (`C`, `X`). Thus, item repetitions are taken into account.

Constraint 5—Regular expression. Such a constraint is satisfied if the pattern is an accepted regular expression as stated by the user. A regular expression can be encoded in ASP as its equivalent deterministic finite automata. Expressing such a constraint is mainly technical and is not detailed here. SPIRIT (Garofalakis et al. 1999) is one of the rare algorithms that considers complex pattern constraints expressed as regular expressions.

Constraint 6—Duration constraints. The duration (or span) of some pattern is the difference between its last item timestamp and its first item timestamp. A duration constraint requires that the pattern duration should be longer or shorter than a given time period. In the database encoding introduced Sect. 4.1, predicate seq(T,P,I) defines the timestamp of I in sequence T as the integer position P. A global constraint such as *max-span* cannot be expressed through simple local constraints on successive pattern item occurrences, as gap constraints described in the next paragraph. In fact, the predicate occS/3 does not describe the embeddings precisely enough to express the *max-span* constraint: for some pattern embedding, there is no explicit link between its first item occurrence and its last item occurrence. The proposed solution is to add an argument to occS/3 to denote the position of the occurrence of the first pattern item:

```
11 %pattern embeddings (skip-gaps strategy)
12 occS(T,1,P,P)        :- seq(T,P,I), pat(1,I).
13 occS(T,Pos+1,P,IP)  :- occS(T,Pos,Q,IP), seq(T,P,I), pat(Pos+1,I)
                         ,
14                        P-IP+1<=maxspan, P-IP+1>=minspan.
```

Constraint 7—Gap constraints. A gap constraint specifies the maximal/minimal number of positions (or timestamp difference) between two successive itemsets in an embedding. The maximal gap constraint is anti-monotonic while the minimal gap is not anti-monotonic. Contrary to pattern constraints, embedding constraints cannot be encoded simply by integrity constraints. In fact, an integrity constraint imposes a constraint on all embeddings. If an embedding does not satisfy the constraint, the whole interpretation—i.e. the pattern—is unsatisfied.

In the following we give an encoding for the *max-gap* and *min-gap* constraints. For such local constraint, the solution consists in modifying the embedding generation (lines 11–12 in Listing 3.2) for yielding only embeddings that satisfy gap constraints:

```
11 occS(T,1,P)        :- seq(T,P,I), pat(1,I).
12 occS(T,Pos+1,P)    :- seq(T,P,I), pat(Pos+1,I), occS(T,Pos,Q),
13                       P-Q-1>=mingap, P-Q-1<=maxgap.
```

This encoding assumes that the value of constants mingap and maxgap have been provided by the user (using #const statements).

Constraints of type 6 and 7 can be mixed by merging the two encodings of occS above:

```
11 occS(T,1,P,P)        :- seq(T,P,I), pat(1,I).
12 occS(T,Pos+1,P,IP)  :- seq(T,P,I), pat(Pos+1,I), occS(T,Pos,Q,IP),
13                        P-Q-1>=mingap, P-Q-1<=maxgap,
14                        P-IP+1<=maxspan, P-IP+1>=minspan.
```

5.2 Condensed Representation of Patterns: Closed and Maximal Sequences

In this section, we study the encodings for two well-studied pattern types, closed and maximal patterns. A closed pattern is such that none of its frequent super-patterns has the same support. A maximal pattern is such that none of its super-patterns is frequent. Thus, it is necessary to compare the supports of several distinct patterns. Since a solution pattern is encoded through an answer set, a simple solution would be to put constraints on sets of answer sets. However, such a facility is not provided by basic ASP language.[4] So, these constraints have been encoded without any comparison of answer sets but as additional constraints on the requested patterns. The next section introduces the definitions of these alternative mining tasks and the properties that were used to transform the pattern set constraints as constraints on individual patterns. Section 5.2.2 gives encodings for closed and maximal patterns extraction.

5.2.1 Definitions and Properties

A frequent pattern S is *maximal* (resp. *backward-maximal*) with respect to the relation $\prec$ (resp. $\prec_b$) iff there is no other frequent pattern S' such that $S \prec S'$ (resp. $S \prec_b S'$).

A frequent pattern S is *closed* (resp. *backward-closed*) with respect to the relation $\prec$ (resp. $\prec_b$) iff there is no proper superpattern S' such that $S \prec S'$ (resp. $S \prec_b S'$) and $supp(S) = supp(S')$. Mining the closed patterns significantly reduces the number of patterns without loss of information for the analyst. Having the closed patterns and their support, the support of any pattern can be computed. This is not the case for maximal patterns.

Example 6 (Maximal and closed-sequential pattern mining) Considering the database of Example 2, among the frequent patterns with $f_{min} = 3$, the only maximal pattern is $\langle abc \rangle$. The set of backward-maximal is $\{\langle c \rangle, \langle bc \rangle, \langle ac \rangle, \langle abc \rangle\}$.

The set of closed patterns is $\{\langle a \rangle, \langle b \rangle, \langle ab \rangle, \langle ac \rangle, \langle abc \rangle\}$. $\langle bc \rangle$ is not closed because in any sequence it occurs, it is preceded by an a. Thus $supp(\langle bc \rangle) = supp(\langle abc \rangle) = 4$.

The set of backward-closed patterns is $\{\langle a \rangle, \langle b \rangle, \langle c \rangle, \langle bc \rangle, \langle ac \rangle, \langle abc \rangle\}$. $\langle bc \rangle$ is backward-closed because any pattern $\langle bc? \rangle$ is frequent.

Now, we introduce alternative maximality/closure conditions. The objective of these equivalent conditions is to define maximality/closure without comparing patterns. Such conditions can be used to encode the mining of condensed pattern representations. The main idea is to say that a sequence S is maximal (resp. closed) if and only if for every sequence S' s.t. S is a subsequence of S' with $|S'| = |S| + 1$, then S' is not frequent (resp. S' has not the same support as S).

More precisely, a frequent pattern S is maximal iff any sequence S_a^j, obtained by adding to S any item a at any position j, is non frequent. Such an a will be called an *insertable* item.

Proposition 3 (Maximality condition) *A frequent sequence $S = \langle t_1, \ldots, t_n \rangle$ is maximal iff $\forall a \in \mathscr{I}, \forall j \in [0, n], |\{T \in \mathscr{D} | S \prec T \wedge S_a^j \prec T\}| < f_{min}$, where $S_a^0 = \langle a, t_1, \ldots, t_n \rangle$, $S_a^j = \langle t_1, \ldots, t_j, a, t_{j+1}, \ldots, t_n \rangle$ and $S_a^n = \langle t_1, \ldots, t_n, a \rangle$.*

---

[4]*asprin* (Brewka et al. 2015) is a *clingo* extension that allows for this kind of comparison. For more details about the use of *asprin* to extract skypatterns, see Gebser et al. (2016).

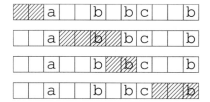

Fig. 3 Illustration of the notion of insertable region on the example of Fig. 2 for pattern $\langle abc \rangle$. Each *line* shows an insertable region, from *top* to *bottom*: insertion in the prefix, insertion between b and b, insertion in b and c, insertion in the suffix

A frequent pattern S is closed iff for any frequent sequence S_a^j, obtained by adding any item a at any position j in S, any sequence T that supports S supports also S_a^j.

Proposition 4 (Closure condition) *A frequent sequence* $S = \langle t_1, \ldots, t_n \rangle$ *is closed iff* $\forall a \in \mathscr{I}, \forall j \in [0, n], supp(S_a^j) \geq f_{min} \implies (\forall T \in \mathscr{D}, S \prec T \implies S_a^j \prec T),$ *where* $S_a^0 = \langle a, t_1, \ldots, t_n \rangle, S_a^j = \langle t_1, \ldots, t_j, a, t_{j+1}, \ldots, t_n \rangle$ *and* $S_a^n = \langle t_1, \ldots, t_n, a \rangle$.

A consequence (the contraposition) of these properties is that if an item may be inserted between items of an embedding for at least f_{min} sequences (resp. for all supported sequences) then the current pattern is not maximal (resp. not closed). The main idea of our encodings is grounded on this observation.

The main difficulty is to construct the set of insertable items for each in-between position of a pattern, so-called insertable regions. Figure 3 illustrates the insertable regions of a sequence for the pattern $\langle abc \rangle$.

Definition 1 (*Insertable item/insertable region*) Let $P = \langle p_i \rangle_{i \in [l]}$ be an l-pattern, $S = \langle s_i \rangle_{i \in [n]}$ be a sequence and $\varepsilon^j = (e_i^j)_{i \in [l]}, j \in [k]$ be the k embeddings of P in S, $k > 0$. An *insertable region* $R_i = [l_i + 1, u_i - 1] \subset [n], i \in [l+1]$ is a set of positions in S where $l_i \overset{def}{=} \min_{j \in [k]} e_{i-1}^j, i \in [2, l+1], u_i \overset{def}{=} \max_{j \in [k]} e_i^j, i \in [1, l]), l_1 \overset{def}{=} 0, u_{l+1} \overset{def}{=} n+1$.

Any item $a \in s_p, p \in [l_i, u_i], i \in [l+1]$ is called an *insertable item* and is such that S supports the pattern P' obtained by inserting a in P at position i as follows:

- $P' = \langle a, p_1, \ldots, p_l \rangle$ if $i = 1$,
- $P' = \langle p_1, \ldots, p_l, a \rangle$ if $i = l+1$,
- $P' = \langle p_1, \ldots, p_{i-1}, a, p_i, \ldots, p_l \rangle$ otherwise.

In the sequel, we present encodings for closed and maximal patterns which are based on the notations introduced in Definition 1. These encodings cope with the most general case of condensed patterns. It should be noted that, for efficiency reasons, most of procedural algorithms for condensed sequential pattern mining process backward-condensed patterns only. Specific ASP encodings for backward-condensed pattern mining can be found in (Guyet et al. 2016). These encodings are known to be more efficient but are less generic. In Sect. 7, the performance of the encodings introduced here will be compared with other existing approaches that often implement only backward closure/maximality constraints.

```
18 % leftmost "valid" embeddings
19 mlocc(T,1,P) :- occS(T,1,P), 0 { occS(T,1,Q): Q<P } 0,
20                  support(T).
21 mlocc(T,X,P) :- occS(T,X,P), mlocc(T,X-1,Q), Q<P, X>1,
22                  support(T).
23
24 % rightmost "valid" embeddings
25 mrocc(T,L,P) :- occS(T,L,P), 0 { occS(T,L,R): R>P } 0, patlen(L).
26 mrocc(T,X,P) :- occS(T,X,P), mrocc(T,X+1,R), R>P, X<L, patlen(L).
27
28 %insertable items
29 ins(T,1  ,I) :- seq(T,P,I), P<Q, mrocc(T,1,  Q).
30 ins(T,X  ,I) :- seq(T,P,I), P<Q, mrocc(T,X,  Q),
31                              P>R, mlocc(T,X-1,R), X>1, patpos(X).
32 ins(T,L+1,I) :- seq(T,P,I), P>R, mlocc(T,L,  R), patlen(L).
```

Listing 3.8 Computation of insertable items—skip-gaps strategy

5.2.2 Encoding Maximal and Closed Patterns Constraints

The encoding below describes how is defined the set of items that can be inserted between successive items of an embedding. These itemsets are encoded by the atoms of predicate ins(T,X,I) where I is an item which can be inserted in an embedding of the current pattern in sequence T between items at position X and X+1 in the pattern. We give the encodings for the two strategies skip-gaps and fill-gaps: Listing 3.8 (resp. Listing 3.9) has to be added to the encoding of skip-gaps strategy (Listing 3.2), resp. fill-gaps strategy (Listing 3.3). We illustrate the way they proceed in Fig. 4.

Listing 3.8 gives the encoding for computing insertable items using the skip-gaps strategy. This encoding is based on the idea that the insertable region i is roughly defined by the first occurrence of the $(i - 1)$-th pattern item and the last occurrence of the i-th pattern item. However, not all occurrences of an item I represented by occS/3 atoms are valid. For instance, in Fig. 4, on the left, the last occurrence of b is not valid because it can not be used to define an occurrence of $\langle abc \rangle$. The valid

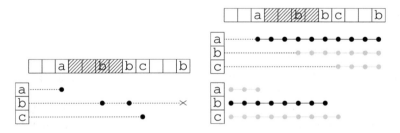

Fig. 4 Illustration of the computation of an insertable region (*hatched area*) where insertable items are located between a and b and related to the first and second element of pattern $\langle abc \rangle$. On the *left*, valid occS/3 atoms in the skip-gaps strategy. In the figures, the *leftmost* occurrences and the *rightmost* occurrences are the same. Concerning the fill-gaps strategy, occF/3 and roccF/3 atoms are illustrated on the *right*. *Black occurrences* are used to compute the hatched region

```
20 %embeddings in a reverse order
21 roccF(T,L,P) :- seq(T,P,I), pat(L,I), patlen(L).
22 roccF(T,L,P) :- roccF(T, L,   P+1), seq(T,P,_).
23 roccF(T,L,P) :- roccF(T, L+1, P+1), seq(T,P,C), pat(L,C).
24
25 %insertable items
26 ins(T,1   ,I) :- seq(T,P,I), roccF(T,1,   P+1).
27 ins(T,L+1,I) :- seq(T,P,I), occF(T,L,P-1), patlen(L).
28 ins(T,X,   I) :- seq(T,P,I), roccF(T,X,P+1),
29                              occF(T,X-1,P-1), patpos(X), X>1.
```

Listing 3.9 Computation of insertable items—fill-gaps strategy

occurrences are those which have both a preceding and a following valid occurrence. Thus, this validity property is recursive. The encoding of Listing 3.8 selects two types of occurrences: the leftmost occurrences (resp. rightmost occurrences) corresponding to the earlier (resp. the later) embeddings.

Lines 19 and 25 are boundary cases. A leftmost occurrence is valid if it is the first occurrence in the sequence. Lines 21–22 expresses that an occurrence of the X-th item is a valid leftmost occurrence if it follows a valid leftmost occurrence of the $(X - 1)$-th item. Note that it is not required to compute a unique leftmost occurrence here. Lines 25–26 do the same operation starting from the end of the sequence, precisely, the rightmost occurrence.

Lines 29–32 define insertable items. There are three cases. Lines 29 and 32 are specific boundary cases, i.e. insertion respectively in the prefix and in the suffix. The rule in lines 30–31 specifies that insertable items I are the items of a sequence T at position P such that P is strictly between a leftmost position of the $(X - 1)$-th item and a rightmost position of the X-th item. In Fig. 4 left, the hatched segment defines the second insertable region for pattern $\langle abc \rangle$ (strictly between a and b).

The encoding of Listing 3.9 achieves the same task using the alternative semantics for predicate occF/3 defining the fill-gaps strategy. As noted for the previous encoding, only the positions of the last and the first valid occurrences are required for any pattern item. It can be noted that the fill-gaps strategy provides the first valid occurrence of an item X as the first atom of the occF(T,X,_) sequence. Then, computing the last occurrence for each pattern item can be done in the same manner considering an embedding represented in reverse order. The right part of Fig. 4 illustrates occF/3 and roccF/3 (reverse order) occurrences (see Listing 3.9, lines 21–23). We can notice that the hatched insertable region is the intersection of occurrences related to a and reverse occurrences related to b, after having removed intersection bounds.

The computation of insertable items, Listing 3.9 lines 26–29, exploits the above remark. Line 26 defines the insertable region in a prefix using roccF(T,1,P). Since items are insertable if they are strictly before the first position, we consider the value of roccF(T,1,P+1). Line 27 uses occF(T,L,P) to identifies the suffix region. Line 28–29 combines both constraints for in-between cases.

We can now define the (integrity) constraints for closed and maximal patterns. These constraints are the same for the two embedding strategies.

To extract only maximal patterns, the following constraint denies patterns for which it is possible to insert an item which will be frequent within sequences that support the current pattern.

```
:- item(I), X = 1..maxlen+1, { ins(T,X,I) : support(T) } >= th.
```

The following constraint concerns the extraction of closed-patterns. It specifies that for each insertion position (from 1, in the prefix, to maxlen+1, in the suffix), it not possible to have a frequent insertable item I for each supported transaction.

```
:- item(I), X = 1..maxlen+1, { ins(T,X,I) } >=th,
                             ins(T,X,I) : support(T).
```

Though interesting from a theoretical point of view, these encodings leads to more complex programs and should be more difficult to ground and to solve, especially the encoding in Listing 3.8. Backward-closure/maximality constraints are more realistic from a practical point of view.

Finally, it is important to notice that condensed constraints have to be carefully combined with other patterns/embedding constraints. As noted by Negrevergne et al. (2013), in such cases the problem is not clearly specified. For instance, with our database of Example 2, extracting closed patterns amongst the patterns of length at most 2 will not yield the same results as extracting closed patterns of length at most 2. In the first case, $\langle bc \rangle$ is closed because there is no extended pattern (of length at most 2) with the same support. In the second case, this pattern is not closed (see Example 6), even if its length is at most 2.

6 Related Works

Sequential pattern mining in a sequence database have been addressed by numerous algorithms inspired by algorithms for mining frequent itemsets. The most known algorithms are GSP (Srikant and Agrawal 1996), SPIRIT (Garofalakis et al. 1999), SPADE (Zaki 2001), PrefixSpan (Pei et al. 2004), and CloSpan (Yan et al. 2003) or BIDE (Wang and Han 2004) for closed sequential patterns. It is worth-noting that all these algorithms are based on the anti-monotonicity property which is essential to obtain good algorithmic performances. The anti-monotonicity property states that if some pattern is frequent then all its sub-patterns are also frequent. And reciprocally, if some pattern is not-frequent then all its super-patterns are non-frequent. This property enables the algorithm to prune efficiently the search space and thus reduces its exploration. These algorithms differ by their strategy for browsing the search space. GSP (Srikant and Agrawal 1996) is based on a breadth-first strategy, while PrefixSpan (Pei et al. 2004) combines a depth-first strategy with a database projection that consists in reducing the database size after each pattern extension. LCM_seq (Uno 2004) is also based on the PrefixSpan principle, but it uses the data structures and processing method of LCM, which is the state of the art algorithm for frequent itemsets mining. Finally, SPADE (Zaki 2001) introduces a vertical representation of database to propose an alternative to the two previous type of algorithms. For more details about these algorithms, we refer the reader to the survey of Mooney and Roddick (2013).

Many algorithms extend the principles of these algorithms to extract alternative forms of sequential patterns. Constraints and condensed patterns are among the most studied alternative patterns due to their relevance to a wide range of applications or to their concise representation of frequent patterns. Integrating constraints in sequential pattern mining is often limited to the use of anti-monotonic temporal constraints such as maxgap constraints. When constraints are not anti-monotonic, the previous pruning technique cannot be applied and the computation may require an exhaustive search, which is not reasonable. The usual technique consists in defining an anti-monotonic upper-bound of the measure such that a large part of the search space can be prune (e.g. high occupancy patterns, Zhang et al. 2015). The tighter the upper-bound is, the better the computing performances are. However, any new type of constraint requires a long effort before being integrated in an efficient algorithm. Integrating flexible and generic constraints in a pattern mining algorithm remains a challenge.

The design of a generic framework for data mining is not a new problem. It has been especially studied within the field of inductive databases as proposed by Imielinski and Mannila (1996). In an inductive database, knowledge discovery is viewed as a querying process. The idea is that queries would return patterns and models. This framework is based on a parallel between database and data mining theory and has as ultimate goal the discovery of a relational algebra for supporting data mining.

In the specific field of pattern mining, designing such query languages has recently attracted interest in the literature (De Raedt 2015; Guns et al. 2015; Negrevergne et al. 2013; Bonchi et al. 2006; Boulicaut and Jeudy 2005; Vautier et al. 2007). For instance, Vautier et al. (2007) proposed a framework which is based on an algebraic specification of pattern mining operators. Bonchi et al. (2006) proposed the ConQueSt system which is an algorithmic framework that accepts constraints with different properties (anti-monotonic, convertible, loose anti-monotonic, etc.). Boulicaut and Jeudy (2005) survey the field of constraint-based data mining. Negrevergne et al. (2013) recently proposed an algebra for programming pattern mining problems. This algebra allows for the generic combination of constraints on individual patterns with dominance relations between patterns.

More recently, the declarative approaches have shown a strong potential to be relevant frameworks for implementing the principles of inductive databases (De Raedt 2015). Many data mining problems can be formalized as combinatorial problems in a declarative way. For instance, tasks such as the discovery of patterns in data, or finding clusters of similar examples in data (Dao et al. 2015), often require constraints to be satisfied and require solutions that are optimal with respect to a given scoring function. The aim of these declarative approaches is to obtain a declarative constraint-based language even at the cost of degraded runtime performance compared to a specialized algorithm. Three types of state-of-the-art solvers have been used: SAT solvers (Coquery et al. 2012), CP solvers (Guns et al. 2015) and ASP solvers (Järvisalo 2011).

MiningZinc (Guns et al. 2015) is a CP-based approach providing a specific language built upon MiniZinc, a medium-level constraint modelling language (Nethercote et al. 2007). A similar declarative language has been proposed by Bruynooghe et al. (2015) using the IDP3 system. IDP3 is a Knowledge Base System (KBS) that intends to offer the user a range of inference methods and to make use of different state of the art technologies including SAT, SAT Modulo Theories, Constraint Programming and various technologies from Logic Programming. One example of application of their system concerns the problem of learning a minimal automaton consistent with a given set of strings. In ASP, Järvisalo (2011) has proposed the first attempt of encoding pattern mining in ASP. Järvisalo addressed this problem as a new challenge for the ASP solver, but he did not highlight the potential benefit of this approach to improve the expressiveness of pattern mining tools. Nonetheless, the first order expressions of ASP encodings can easily be understood by users without higher abstracted languages. Following Guns et al. 's proposal (Guns et al. 2011), Järvisalo designed an ASP program to extract frequent itemsets in a transaction database. A major feature of Järvisalo's proposal is that each answer set (AS) contains only one frequent itemset associated with the identifiers of the transactions where it occurs. To the best of our knowledge, there is no comprehensive language provided for SAT-based data mining approaches.

All these approaches were conducted on itemset mining in transaction databases, which is much simpler than sequential pattern mining in a sequence database. Some recent works have proposed to explore declarative programming for sequential pattern mining. In fact, dealing with expressive constraints is especially interesting for sequential pattern mining. The range of constraints on sequential patterns is wider than on itemsets and are meaningful for various concrete data analysis issues.

Negrevergne and Guns (2015) proposed the CPSM approach which can be considered as the state of the art of declarative sequential pattern mining. Their contribution is twofold: (i) the first declarative encodings of the standard sequential pattern mining task, (ii) an efficient CP-based approach based on dedicated propagators that remains compatible with sequential pattern constraints. By combining efficiency and declarativity, CPSM is a proof of concept that a declarative approach can be efficient to solve pattern mining tasks.

Métivier et al. (2013) have developed a constraint programming method for mining sequential patterns with constraints in a sequence database. The constraints are based on *amongst* and *regular expression* constraints and expressed by automata. Coquery et al. (2012) have proposed a SAT based approach for sequential pattern mining. The patterns are of the form *ab?c* and an occurrence corresponds to an exact substring (without gap) with joker (the character ? replaces exactly one item different from *b* and *c*). Coletta and Negrevergne (2016) have proposed a purely boolean SAT formulation of sequential pattern mining (including closed and maximal patterns) that can be easily extended with additional constraints.

ASP has also been used for sequential pattern mining (Gebser et al. 2016; Guyet et al. 2014). Gebser et al. (2016) have proposed, firstly, an efficient encoding for sequential pattern mining. Secondly, they have proposed to use the *asprin* system for the management of pattern set constraints using preferences. In Guyet et al. (2014), the mining task is the extraction of serial episodes in a unique long sequence of itemsets where occurrences are the minimal occurrences with constraints. Counting the number of occurrences of a pattern, or of a set of patterns, in a long sequence introduces additional complexity compared to mining sequential patterns from a sequence database since two pattern occurrences can overlap. The main contribution is a method for enumerating pattern occurrences that ensures the anti-monotonicity property.

7 Experiments

Having demonstrated that modelling in ASP is powerful yet simple, it is now interesting to examine the computational behavior of ASP-based encodings.

The first experiments compare the performance, in runtime and memory requirements, of the various ASP programs presented before. The objective is to better understand the advantages and drawbacks of each encoding. The questions we would like to answer are: which of the two embedding strategies is the best? does the encoding improvement really reduce computing resources needs? what is the behaviour of our encoding with added pattern constraints?

Next, we compare our results with the CP-based ones of CPSM (Negrevergne and Guns 2015). CPSM constitutes a natural reference since it aims at solving a mining task similar to the present one and since CPSM adopts a semi-declarative approach, in particular, occurrence search is performed by a dedicated constraint propagator.

In all presented experiments, we use the version 4.5 of *clingo*,[5] with default solving parameters. For benchmarking on synthetic data, the ASP programs were run on a computing server with 8Go RAM without using the multi-threading mode of clingo. Multi-threading reduces the mean runtime but introduces variance due to the random allocation of tasks. Such variance is inconvenient for interpreting results with repeated executions. For real datasets, we used the multi-threading mode with 4 threads and 20Go shared RAM. This large amount of memory is required for large datasets.

7.1 Encodings Comparisons on Synthetic Datasets

The first experiments were conducted on synthetic databases to control the most important features of data. It allows for an easier and more reliable analysis of time and memory requirements with respect to these parameters. We designed a sequential database simulator to generate datasets with controlled characteristics. The

---

[5]https://potassco.org/.

Table 1 Sequence generator parameters

Parameter	Default value	Description
D	500	Number of sequences in the database
l	20	Sequence mean length (sequence length follows a normal law)
n	20	Number of different patterns
lp	5	Pattern mean length
th_D	10%	Minimum number of occurrences generated for each pattern
k	50	Alphabet size. The distribution of item occurrences follows a normal law ($\mu = 0.5$ and $\sigma = 0.05$). Some items occur more often than others

generator[6] is based on a "retro-engineering" process: (1) a set of random patterns is generated, (2) occurrences of patterns are assigned to a given percentage of database sequences, and (3) each sequence of items is randomly generated according to the patterns it must contain and a mean length.

The parameters of the generator and their default values are sum up in Table 1. Default values are those used when not explicitly specified.

The task to be solved is the extraction of the complete set of frequent patterns (see Sect. 3). It should be noted that every encoding extracts exactly the same set of patterns. Resource requirements are thus fairly comparable. The computation runtime is the time needed to extract all the patterns. It includes both grounding and solving of the ASP programs using the quiet *clingo* mode (no printed output). The memory consumption is evaluated from the size of the grounded program, i.e. the number of grounded atoms and rules. This approximation is accurate to compare ASP encodings. The solving process may require additional memory. This memory requirement is negligible compared to grounding.

We start with an overall comparison of the different encodings and their refinements with respect to parameters th_D and l. Figure 5 compares the runtimes for different encodings and the two embedding strategies, fill-gaps and skip-gaps. For each setting, 6 databases with the same characteristics were generated. Figure curves show the mean rutime of the successful executions, i.e. those that extract the complete set of frequent pattern within the timeout period. The timeout was set to 20 min.

The exponential growth of the runtime when the threshold decreases is a classical result in pattern mining considering that the number of patterns grows exponentially. Every approach conforms to this behaviour. In more details:

- the longer the sequences, the greater the runtime. Most problem instances related to databases with $l = 10$ can be solved by any approach. When the mean length of sequences increases, the computation time increases also and the number of instances solved within the timeout period decreases. This can be easily explained

---

[6]The generator and databases used in our experiments are available at https://sites.google.com/site/aspseqmining.

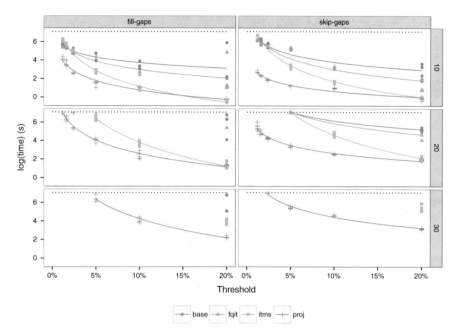

Fig. 5 Mean computation time for synthetic databases with sequences length from 10 to 30 (in *rows*). Each *curve* represents an improvement to the basic encoding: none (base), frequent items (fqit), itemsets (itms), projection (proj). Each *dot* represents the mean of results on 6 datasets. The *left-hand* (resp. *right-hand*) *column* gives the results for the fill-gaps (resp. skip-gaps) strategy. The *dashed horizontal line* denotes the timeout of 20 min

by the combinatorics of computing embeddings which increases with the sequence length.

- all proposed improvements do improve runtime for high frequency thresholds on these small synthetic databases. For $f_{min} = 20\%$, the curve of every proposed improvement is below the curve of the basic encoding. For high thresholds, prefix-projection and itemsets improvements are significantly better. Nonetheless, the lower the threshold, the lower the difference between computation times. This shows that, except for prefix-projection, the improvements are not so efficient for hard mining tasks.
- the prefix-projection improvement is the fastest and reduces significantly the computation time (by 2–3 orders of magnitude).
- the skip-gaps strategy is more efficient than the fill-gaps strategy for these small datasets. The skip-gaps strategy requires less time to extract the same set of patterns than the fill-gaps strategy, for the same encoding improvements.

We will see below that this last result does not accurately predict which strategy should be preferred for mining real datasets. Before, we analyse the memory requirements of the different encodings.

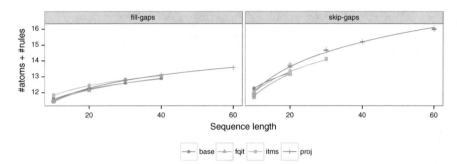

Fig. 6 Memory requirement with respect to sequence length. Problem size (estimated memory requirement) for synthetic databases of sequences of length from 10 to 30 (in *rows*), under frequency threshold of 20%

We first note that the memory consumption is not related to the frequency threshold. This is a specificity of declarative pattern mining. Thus, Fig. 6 compares the embedding strategies only for a unique frequency threshold $f_{min} = 20\%$. The curves show the number of grounded atoms and rules. As it represents a tight approximation of the memory requirement, we will refer to memory in the sequel.

Unsurprisingly, the richer the encoding is, the more memory is required. But the differences are not really significant, except for the prefix-projection programs (proj) which requires the highest number of atoms. We can see that using frequent itemsets (itms) is efficient to reduce the memory requirement. This means that the grounding step was able to exploit the additional rules to avoid the creation of useless atoms and rules. Such a kind of rules is really interesting because, as the algorithmic complexity of the mining task is not high, the efficiency of the ASP program is related to his grounding size.

In addition, from this last point of view, we can note that the fill-gaps strategy requires several order less memory than the skip-gaps strategy. The longer the sequences, the larger the difference. This result is illustrated by Fig. 7. For each problem instance, the ratio of memory usage is computed by dividing the memory required by encoding with skip-gaps strategy with the memory required by the similar encoding with the fill-gaps strategy. Figure 7 illustrates with boxplots the dispersion of these ratios for different sequence lengths. Figure 7 clearly shows that the longer the sequences are, the more efficient the fill-gaps strategy is for memory consumption.

To end this overall comparison, it is interesting to come back to runtime. The overall results of Fig. 5 show that the skip-gaps strategy seems better, but considering that the fill-gaps strategy requires less memory, it is interesting to analyse the evolution of computation time with respect to database size.

Figure 8 illustrates the ratio of runtimes in both strategies when the database size increases. The support threshold, f_{min}, is fixed to 10% and the sequence mean length to 20. We used the prefix-projection encoding for this experiment. Similarly to the previous figure, the ratios were individually computed for each pair of results (fill-gaps/skip-gaps) and the figure shows statistics about these ratio.

Fig. 7 Dispersion of ratios of memory consumption obtained for the skip-gaps strategy to those obtained for the fill-gaps strategy. *Boxplots* were computed for problem instances with threshold at 20% and for all lengths and all encodings

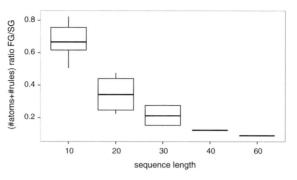

Fig. 8 Dispersion of ratios of runtime of the skip-gaps strategy to the runtime of the fill-gaps strategy

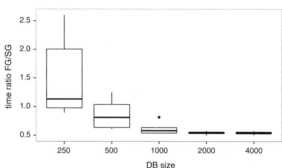

Figure 8 shows clearly that when the database size increases, the fill-gaps strategy becomes more efficient than the skip-gaps strategy.

From these experiments, we can conclude that combining prefix-projection with the fill-gaps strategy gives the best encoding. Thus, in the next section, we will compare this encoding with CPSM.

7.2 Real Dataset Analysis

In these experiments, we analyse the proposed encodings on processing real datasets. We use the same real datasets as selected in Negrevergne and Guns (2015) to have a representative panel of application domains:

- JMLR: a natural language processing dataset; each transaction is a paper abstract from the Journal of Machine Learning Research,
- UNIX: each transaction is a series of shell commands executed by a user during one session,
- iPGR: each transaction is a sequence of peptides that is known to cleave in presence of a Trypsin enzyme,
- FIFA: each transaction is a sequence of webpages visited by a user during a single session.

Table 2 Dataset characteristics: alphabet size, number of sequences and items, max and mean length of sequences, dataset density

Dataset	$|\mathscr{I}|$	$|D|$	$||D||$	$max|T|$	$avg|T|$	Density
Unix user	265	484	10935	1256	22.59	0.085
JMLR	3847	788	75646	231	96.00	0.025
iPRG	21	7573	98163	13	12.96	0.617
FIFA	20450	2990	741092	100	36.239	0.012

The dataset characteristics are sum up in Table 2. Some of them are similar to those of simulated datasets.

7.2.1 Comparison of Frequent Pattern Mining with CPSM

Figure 9 compares the runtimes of ASP-based sequence mining (using the ASP system *clingo*) and CPSM (based on the CP solver *gecode*). We ran the two versions of CPSM. CPSM makes use of global constraints to compute embeddings. This version is known to be very efficient, but it cannot cope with embedding constraints, while CPSM-emb does but is less efficient. We do not compare our approach with dedicated algorithms, which are known to be more efficient than declarative mining approaches (see Negrevergne and Guns 2015 for such comparisons). The timeout was set to 1 h.

The results show that the runtimes obtained with *clingo* are comparable to CPSM-emb. It is lower for IPGR, very similar for UNIX and larger for JMLR. These results are consistent to those presented in Gebser et al. (2016) for synthetic datasets. When sequences become large, the efficiency of our encoding decreases somewhat. The mean length for JMLR is 96 while it is only 12.96 for iPRG. For CPSM with global constraints, the runtime-efficiency is several order of magnitude faster. To be fair, it should be noted that ASP approach ran with four parallel threads while CPSM-emb ran with no multi-threading since it does not support it. It should also be noted that CPSM requires a lot of memory, similarly to ASP-based solving.

7.2.2 Comparison of Constrained Frequent Pattern Mining with CPSM

In this section, we detail the performance on constrained pattern mining tasks. We compare our approach with CPSM-emb, which enables *max-gap* and *max-span* constraints. In this experiments we took the same setting as the experiments of Negrevergne and Guns (2015): we add first a constraint *max-gap* = 2 and then we combine it with a second constraint *max-span* = 10. For each setting, we compute the frequent patterns with our ASP encoding and with CPSM for the four datasets.

Figure 10 shows the runtime and the number of patterns for each experiment. This figure illustrates results for completed searches. A first general remark is that adding

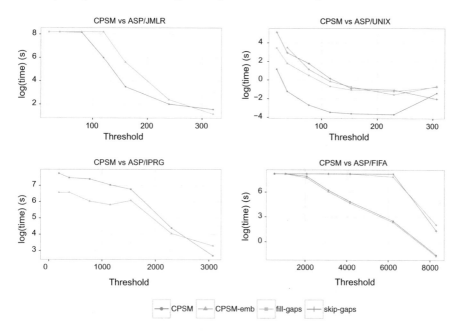

Fig. 9 Runtime for mining frequent patterns with four approaches: CPSM, CPSM-emb, ASP with fill-gaps, ASP with skip-gaps

constraints to ASP encodings reduces computation times. Surprisingly for CPSM, for some thresholds the computation with some constraints requires more time than without constraints. This is the case for example for the iPRG dataset: CPSM could not solve the mining problem within the timeout period for thresholds 769 and 384. Surprisingly, it could complete the task for lower thresholds whereas the task should be more difficult. ASP required also more time to solve the same problem instances, but it could complete them. Again, we can note that the mean sequence length impacts the performance of ASP encodings. CPSM has lower runtime on JMLR than ASP while it is the opposite on iPRG.

The curves related to the number of patterns demonstrate that the number of extracted pattern decreases when the number of constraints increases. Since we present only the results of completed solving, CPSM and ASP yield the same set of patterns.

7.2.3 Analysis of Condensed Pattern Extraction

Figure 11 illustrates the results for condensed pattern mining. This approach cannot be compared to CPSM since it does not propose means for encoding such kind of patterns.

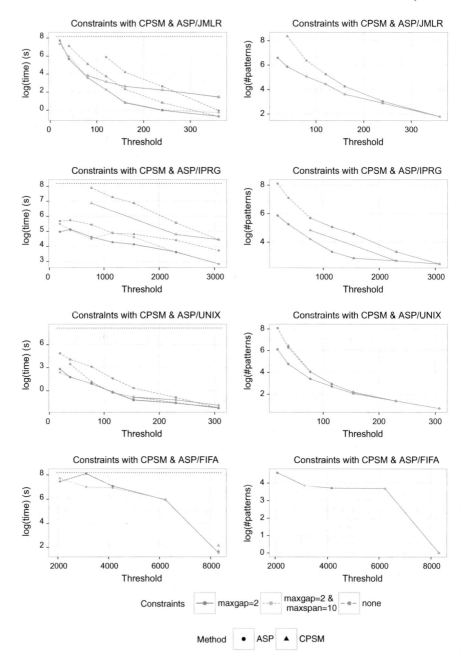

Fig. 10 Results for constrained sequence mining tasks with ASP versus CPSM. On the *left* runtime; on the *right* number of patterns. The two figures are in log scales. From *top* to *bottom*, JMLR, IPRG, UNIX and FIFA. For each plot, the *curves* illustrate the results for different type of constraints (see legend). The *horizontal dashed line* figures out the 1 h timeout

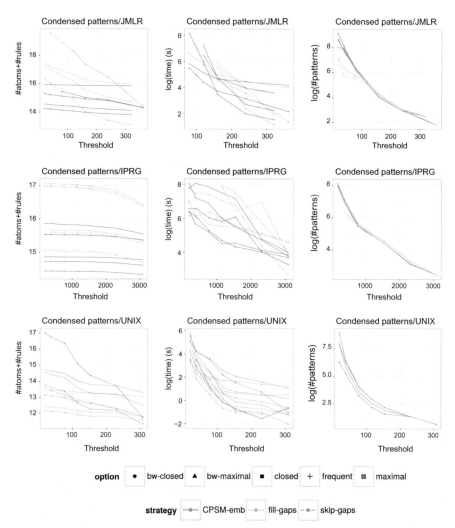

Fig. 11 From *left* to *right*, problem size, runtime and number of extracted patterns with respect to the frequency threshold. Runtimes are shown only if the solving was complete, contrary to pattern numbers which show the number of extracted patterns within the timeout period. From *top* to *bottom*, JMLR, IPRG and UNIX. For each plot, the *curves* illustrate the results for different types of condensed patterns (see legend) and for the two embedding strategies (*fill-gaps* in *red-plain line*, *skip-gaps* in *blue-dashed line*)

This experiment compares the resource requirements in time and memory for mining closed/maximal and backward-closed/maximal patterns. For each of these mining task, we compared the skip-gaps and fill-gaps strategies. The main encoding is still based on prefix-projection. Three real datasets have been processed (JMLR, UNIX and IPRG). The FIFA dataset was not processed due to its heavy memory requirement for some of these tasks.

We can first note that the difference between the number of extracted patterns is low. As expected, all encodings that complete a given mining task extract the same number of patterns. This result supports the correctness of our approach. From the memory point of view, we see that the encodings extracting condensed patterns requires several order of magnitude more memory, especially for (backward-)closed patterns. It is also interesting to note that the memory requirement for the fill-gaps strategy is not linked to the threshold, contrary to the skip-gaps strategy. Again, the fill-gaps strategy seems to be more convenient for small thresholds. We can note that there is a big difference between datasets concerning runtime. For instance, frequent patterns are faster to extract for JMLR and UNIX, but maximal patterns are faster to compute on IPRG. The density of this last dataset makes maximal pattern extraction easier. Uniformly, we can conclude that fill-gaps is faster than skip-gaps. The complexity of the encoding related to insertable items with skip-gaps makes the problem difficult to solve. Opposed to the experiments presented in Guyet et al. (2016), we did not use any solving heuristic. For maximal patterns, a huge improvement of runtime was observed when using the *subset-minimal* heuristic.[7]

8   Conclusion and Perspectives

This article has presented a declarative approach of sequential pattern mining based on answer set programming. We have illustrated how to encode a broad range of mining tasks (including condensed representations and constrained patterns) in pure ASP. Thus, we shown the first advantage of declarative pattern mining: for most well-specified tasks, the development effort is significantly lower than for procedural approaches. The integration of new constraints within our framework requires only few lines of code. This was made possible thanks to the flexibility of both ASP language and solvers.

Nonetheless, another objective of this paper was to give the intuition to the reader that while encoding a straightforward solution to a problem can be easy in ASP, writing efficient programs may be complex. Developing competitive encodings requires a good understanding of the solving process. To this end, we have presented several possible improvements of basic sequential pattern mining and two alternatives for encoding the main complex task, i.e. computing embeddings. These encodings have been extensively evaluated on synthetic and real datasets to draw conclusions about the overall efficiency of this approach (especially compared to the constraint programming approach CPSM) and about which are the best encodings among the proposed ones and in which context.

The first conclusion of these experiments is that our ASP approach has comparable computing performances with CPSM-emb as long as the length of the sequence remains reasonable. This can be explained considering that solving the embedding problem is a difficult task for pure declarative encodings while CPSM relies on

---

[7]The use of subset-minimal heuristic keeps solving the maximal patterns problem complete.

dedicated propagators. The propagators of CPSM solve the embedding problem using additional procedural code. It turns that, for solving the embedding problem in ASP, encoding using a *fill-gaps* strategy appears to be better than using the *skip-gaps* strategy on real datasets, thanks to lower memory requirements.

The second conclusion is that adding constraints on patterns reduces runtime, but increases memory consumption. For real datasets, the more constraints are added, the more memory is required. This is due, to encoding the constraints, but also to encoding the information that may be required to compute constraints. For example, encodings using the *maxspan* constraint require more complex embeddings (occS/4 atoms) than encodings without this constraint.

To fully benefit from the flexibility of our approach to proceed large datasets, we need to improve the efficiency of the computation of embeddings. Our objective is now to mimic the approach of CPSM consisting in using propagators within the solver to solve the part of the problems for which procedural approaches are efficient. The new *clingo* 5 series will integrate "ASP modulo theory" solving processes. This new facilities will enable to combine ASP and propagators in an efficient way.

Acknowledgements We would like to thanks Roland Kaminski and Max Ostrowski for their significant inputs and comments about ASP encodings; and Benjamin Negrevergne and Tias Guns for their suggestions about the experimental part. We also thank the anonymous reviewers for their valuable comments and constructive suggestions.

Appendix

Listing 3.10 illustrates how the encoding of the *skip-gaps* strategy can be transformed to mine sequential patterns that are sequences of itemsets.

```
1 item(I) :- seq(_, _,I). %set of items
2
3 %sequential pattern generation
4 patpos(1).
5 0 { patpos(X+1) } 1 :- patpos(X), X<maxlen.
6 patlen(L) :- patpos(L), not patpos(L+1).
7
8 %generate an itemset for each position
9 1 { pat(X,I): item(I) } :- patpos(X).
10
11 %pattern occurrences
12 occS(T,1,P) :- seq(T,P,I):pat(1,I); seq(T,P,_).
13 occS(T,Pos+1,P) :- occS(T,Pos,Q), Q<P, seq(T,P,J),
14                    pat(Pos+1,J), seq(T,P,I):pat(Pos+1,I).
15
16 support(T) :- occS(T, L, _), patlen(L).
17 :- { support(T) } th-1.
```

Listing 3.10 Mining sequences with patterns as sequences of itemsets

The first difference with the encoding of Listing 3.2 concerns the generation of patterns. The upper bound constraint of the choice rule in Line 9 has been removed, enabling the possible generation of every non-empty subset of $\mathscr{I}$.

The second difference is that the new ASP rules verify the inclusion of all items in itemsets. Line 14, `seq(T,P,I):pat(1,I)` indicates that for each atom `pat(1,I)` there should exist an atom `seq(T,P,I)` to satisfy the rule body. A similar expression is used Line 15.

References

Agrawal, R., Imielinski, T., & Swami, A. (1993). Mining association rules between sets of items in large databases. In *Proceedings of the ACM SIGMOD Conference on Management of Data* (pp. 207–216).

Agrawal, R., & Srikant, R. (1995). Mining sequential patterns. In *Proceedings of the International Conference on Data Engineering* (pp. 3–14).

Biere, A., Heule, M., van Maaren, H., & Walsh, T. (2009). *Handbook of satisfiability. Frontiers in artificial intelligence and applications* (Vol. 185). IOS Press.

Bonchi, F., Giannotti, F., Lucchese, C., Orlando, S., Perego, R., & Trasarti, R. (2006). Conquest: A constraint-based querying system for exploratory pattern discovery. In *Proceedings of the International Conference on Data Engineering* (pp. 159–159).

Boulicaut, J.-F., & Jeudy, B. (2005). Constraint-based data mining. In O. Maimon & L. Rokach (Eds.), *Data mining and knowledge discovery handbook* (pp. 399–416). US: Springer.

Brewka, G., Delgrande, J.P., Romero, J., & Schaub, T. (2015). Asprin: Customizing answer set preferences without a headache. In *Proceedings of the Conference on Artificial Intelligence (AAAI)*, pp. 1467–1474.

Bruynooghe, M., Blockeel, H., Bogaerts, B., De Cat, B., De Pooter, S., Jansen, J., et al. (2015). Predicate logic as a modeling language: Modeling and solving some machine learning and data mining problems with IDP3. *Theory and Practice of Logic Programming, 15*(06), 783–817.

Coletta, R., & Negrevergne, B. (2016). A SAT model to mine flexible sequences in transactional datasets. arXiv:1604.00300.

Coquery, E., Jabbour, S., Saïs, L., & Salhi, Y. (2012). A SAT-Based approach for discovering frequent, closed and maximal patterns in a sequence. In *Proceedings of European Conference on Artificial Intelligence (ECAI)* (pp. 258–263).

Dao, T., Duong, K., & Vrain, C. (2015). Constrained minimum sum of squares clustering by constraint programming. In *Proceedings of Principles and Practice of Constraint Programming* (pp. 557–573).

De Raedt, L. (2015). Languages for learning and mining. In *Proceedings of the Conference on Artificial Intelligence (AAAI)* (pp. 4107–4111).

Garofalakis, M., Rastogi, R., & Shim, K. (1999). SPIRIT: Sequential pattern mining with regular expression constraints. In *Proceedings of the International Conference on Very Large Data Bases* (pp. 223–234).

Gebser, M., Guyet, T., Quiniou, R., Romero, J., & Schaub, T. (2016). Knowledge-based sequence mining with ASP. In *Proceedings of International Join Conference on Artificial Intelligence* (pp. 1497–1504).

Gebser, M., Kaminski, R., Kaufmann, B., Ostrowski, M., Schaub, T., & Schneider, M. (2011). Potassco: The Potsdam answer set solving collection. *AI Communications, 24*(2), 107–124.

Gebser, M., Kaminski, R., Kaufmann, B., Schaub, T. (2014). *Clingo = ASP + control: Preliminary report*. In *Technical Communications of the Thirtieth International Conference on Logic Programming*.

Gelfond, M., & Lifschitz, V. (1991). Classical negation in logic programs and disjunctive databases. *New Generation Computing, 9*, 365–385.

Guns, T., Dries, A., Nijssen, S., Tack, G., & De Raedt, L. (2015). MiningZinc: A declarative framework for constraint-based mining. *Artificial Intelligence*, page In press.

Guns, T., Nijssen, S., & De Raedt, L. (2011). Itemset mining: A constraint programming perspective. *Artificial Intelligence, 175*(12–13), 1951–1983.

Gupta, M., & Han, J. (2013). *Data mining: Concepts, methodologies, tools, and applications*, chapter Applications of pattern discovery using sequential data mining (pp. 947–970). IGI-Global.

Guyet, T., Moinard, Y., & Quiniou, R. (2014). Using answer set programming for pattern mining. In *Proceedings of Conference "Intelligence Artificielle Fondamentale" (IAF)*.

Guyet, T., Moinard, Y., Quiniou, R., & Schaub, T. (2016). Fouille de motifs séquentiels avec ASP. In *Proceedings of Conference "Extraction et la Gestion des Connaissances" (EGC)* (pp. 39–50).

Imielinski, T., & Mannila, H. (1996). A database perspective on knowledge discovery. *Communications of the ACM, 39*(11), 58–64.

Janhunen, T., & Niemelä, I. (2016). The answer set programming paradigm. *AI Magazine, 37*, 13–24.

Järvisalo, M. (2011). Itemset mining as a challenge application for answer set enumeration. In *Proceedings of the Conference on Logic Programming and Nonmonotonic Reasoning* (pp. 304–310).

Lallouet, A., Moinard, Y., Nicolas, P., & Stéphan, I. (2013). Programmation logique. In P. Marquis, O. Papini, & H. Prade (Eds.), *Panorama de l'intelligence artificielle: ses bases méthodologiques, ses développements* (Vol. 2). Cépaduès.

Lefèvre, C., & Nicolas, P. (2009). The first version of a new ASP solver: ASPeRiX. In *Proceedings of the Conference on Logic Programming and Nonmonotonic Reasoning* (pp. 522–527).

Leone, N., Pfeifer, G., Faber, W., Eiter, T., Gottlob, G., Perri, S., et al. (2006). The DLV system for knowledge representation and reasoning. *ACM Transactions on Computational Logic, 7*(3), 499–562.

Lhote, L. (2010). Number of frequent patterns in random databases. In Skiadas, C. H. (Ed.), *Advances in data analysis*, Statistics for industry and technology (pp. 33–45).

Lifschitz, V. (2008). What is answer set programming? In *Proceedings of the Conference on Artificial Intelligence (AAAI)* (pp. 1594–1597).

Low-Kam, C., Raïssi, C., Kaytoue, M., & Pei, J. (2013). Mining statistically significant sequential patterns. In *Proceedings of the IEEE International Conference on Data Mining* (pp. 488–497).

Métivier, J.-P., Loudni, S., & Charnois, T. (2013). A constraint programming approach for mining sequential patterns in a sequence database. In *Proceedings of the Workshops of the European Conference on Machine Learning and Principles and Practice of Knowledge Discovery in Databases (ECML/PKDD)*.

Mooney, C. H., & Roddick, J. F. (2013). Sequential pattern mining—Approaches and algorithms. *ACM Computing Surveys, 45*(2), 1–39.

Muggleton, S., & De Raedt, L. (1994). Inductive logic programming: Theory and methods. *The Journal of Logic Programming, 19*, 629–679.

Negrevergne, B., Dries, A., Guns, T., & Nijssen, S. (2013). Dominance programming for itemset mining. In *Proceedings of the International Conference on Data Mining* (pp. 557–566).

Negrevergne, B., & Guns, T. (2015). Constraint-based sequence mining using constraint programming. In *Proceedings of International Conference on Integration of AI and OR Techniques in Constraint Programming, CPAIOR* (pp. 288–305).

Nethercote, N., Stuckey, P. J., Becket, R., Brand, S., Duck, G. J., & Tack, G. (2007). MiniZinc: Towards a standard CP modelling language. In *Proceedings of the Conference on Principles and Practice of Constraint Programming* (pp. 529–543).

Pei, J., Han, J., Mortazavi-Asl, B., Wang, J., Pinto, H., Chen, Q., et al. (2004). Mining sequential patterns by pattern-growth: The prefixspan approach. *IEEE Transactions on Knowledge and Data Engineering, 16*(11), 1424–1440.

Pei, J., Han, J., & Wang, W. (2007). Constraint-based sequential pattern mining: The pattern-growth methods. *Journal of Intelligent Information Systems, 28*(2), 133–160.

Perer, A., & Wang, F. (2014). Frequence: Interactive mining and visualization of temporal frequent event sequences. In *Proceedings of the international Conference on Intelligent User Interfaces* (pp. 153–162).

Rossi, F., Van Beek, P., & Walsh, T. (2006). *Handbook of constraint programming*. Elsevier.

Shen, W., Wang, J., & Han, J. (2014). Sequential pattern mining. In Aggarwal, C. C., & Han, J. (Ed.), *Frequent pattern mining* (pp. 261–282). Springer.

Simons, P., Niemelä, I., & Soininen, T. (2002). Extending and implementing the stable model semantics. *Artificial Intelligence, 138*(1–2), 181–234.

Srikant, R., & Agrawal, R. (1996). Mining sequential patterns: Generalizations and performance improvements. In *Proceedings of the 5th International Conference on Extending Database Technology* (pp. 3–17).

Syrjänen, T., & Niemelä, I. (2001). The smodels system. In *Proceedings of the Conference on Logic Programming and Nonmotonic Reasoning* (pp. 434–438).

Ugarte, W., Boizumault, P., Crémilleux, B., Lepailleur, A., Loudni, S., Plantevit, M., Raïssi, C., & Soulet, A. (2015). Skypattern mining: From pattern condensed representations to dynamic constraint satisfaction problems. *Artificial Intelligence*, page In press.

Uno, T. (2004). http://research.nii.ac.jp/~uno/code/lcm_seq.html.

Vautier, A., Cordier, M., & Quiniou, R. (2007). Towards data mining without information on knowledge structure. In *Proceedings of the Conference on Principles and Practice of Knowledge Discovery in Databases* (pp. 300–311).

Wang, J., & Han, J. (2004). BIDE: Efficient mining of frequent closed sequences. In *Proceedings of the International Conference on Data Engineering* (pp. 79–90).

Yan, X., Han, J., & Afshar, R. (2003). CloSpan: Mining closed sequential patterns in large datasets. In *Proceedings of the SIAM Conference on Data Mining* (pp. 166–177).

Zaki, M. J. (2001). SPADE: An efficient algorithm for mining frequent sequences. *Journal of Machine Learning, 42*(1/2), 31–60.

Zhang, L., Luo, P., Tang, L., Chen, E., Liu, Q., Wang, M., et al. (2015). Occupancy-based frequent pattern mining. *ACM Transactions on Knowledge Discovery from Data, 10*(2), 1–33.

Author Biographies

Thomas Guyet is assistant professor at AGROCAMPUS-OUEST and he is doing his research in the Inria/IRISA LACODAM Team. The research interests of Thomas Guyet range from cognitive foundations to practical application of discovering spatial and temporal patterns in semantically complex datasets. He develops research in large range of artificial intelligence domains including (sequential) pattern mining, knowledge dicovery, declarative programming (answer set programming).

Yves Moinard was research fellow at Inria. His research interest is the foundation of knowledge reasoning. He more especially contributed to the theorical frameworks of default reasoning and circunscription which are foundations of ASP. He recently was interested in causal reasoning and practical aspects of ASP.

René Quiniou is research fellow at Inria. His research interests include machine learning, data mining, model-based diagnosis and monitoring. He has a special interest in temporal pattern discovery from streaming observation data coming from dynamic systems, e.g. numerical data recorded by sensors or timestamped symbolic events recorded in online logs.

Torsten Schaub is university professor at the University of Potsdam, Germany, and holds an international chair at Inria Rennes, France. He is a fellow of ECCAI and the current president of the Association of Logic Programming. His current research focus lies on Answer set programming (ASP) and its applications, which materializes at http://www.potassco.org, the home of the open source project Potassco bundling software for ASP developed at Potsdam.

Consistency-Latency Trade-Off of the LibRe Protocol: A Detailed Study

Sathiya Prabhu Kumar, Sylvain Lefebvre, Raja Chiky
and Olivier Hermant

Abstract In multi-writer, multi-reader systems, data consistency is ensured by the number of replica nodes contacted during read and write operations. Contacting a sufficient number of nodes in order to ensure data consistency comes with a communication cost and a risk to data availability. In this paper, we extend our previous work on a consistency protocol called LibRe, which helps to read the latest version of a data item by contacting a minimum number of replica nodes. The protocol uses a registry that records the set of replica nodes containing the most recent version of the data items until all replicas of this data item converge to a consistent state. Hence, referring to the registry during read time helps to forward the read requests to the replica nodes holding the most recent version of the needed data item. In the following work, we show that this protocol provides a new trade-off between consistency and latency for distributed data storage systems. We provide a formal description of the protocol and its reliability and evaluate the scalability of the protocol up to one hundred nodes by simulation. We also demonstrate the effectiveness of the approach in practice by providing a proof-of-concept implementation of the protocol inside the Cassandra distributed data store. The test results prove that using LibRe protocol, an application would experience a similar number of stale reads compared to strong consistency options offered by Cassandra, while achieving lower latency, and similar availability.

S.P. Kumar (✉) · S. Lefebvre · R. Chiky
ISEP, 28, rue Notre-Dame des Champs, 75006 Paris, France
e-mail: sathiya-prabhu.kumar@isep.fr

S. Lefebvre
e-mail: Sylvain.Lefebvre@isep.fr

R. Chiky
e-mail: Raja.Chiky@isep.fr

O. Hermant
MINES ParisTech, PSL Research University, 60 Bd St-Michel, 75006 Paris, France
e-mail: Olivier.Hermant@mines-paristech.fr

© Springer International Publishing AG 2018   83
B. Pinaud et al. (eds.), *Advances in Knowledge Discovery and Management*,
Studies in Computational Intelligence 732, https://doi.org/10.1007/978-3-319-65406-5_4

1   Introduction

In the wake of the Big Data movement and the emergence of planet scale content sharing systems, distributed storage has become an important and critical part of many software systems. In particular, distributed database systems aim at ensuring the scalability and reliability of the data layer by copying data on several, possibly remote nodes for both performance and availability in case of hardware failure. This copying process is called replication, and each piece of copied data is called a replica. Although a data is physically copied to more than one nodes in the system, they are logically the same. Thus, any change in one of the replicas has to be reflected on the other replicas in a timely manner. If all replicas are in the same state, then the data is said to be consistent. However, ensuring data consistency with high availability and minimum request latency is notoriously challenging (Abadi 2012; Gilbert and Lynch 2002).

Quorum-based replication systems use the so-called intersection property (Naor and Wool 1998; Vukolic 2010) to guarantee an efficient and consistent view of the replicas to all users accessing the data. For example the Dynamo (DeCandia et al. 2007) distributed key-value store built by Amazon, and similar systems such as Cassandra (Lakshman and Malik 2010), Voldemort (Voldemort 2015) and Riak (Klophaus 2010) follow this model. This intersection property can be expressed by the formula $R + W > N$. This formula symbolizes that the system ensures consistency if the sum of the nodes acknowledging the Read operations (R) and the nodes acknowledging the Write operations (W) is greater than the total number of replica nodes (N). If the intersection property cannot be satisfied, the system will reject the operation.

Satisfying the intersection property requires contacting a majority of nodes during read and/or write operations. This approach penalizes the request latency and adds threat to data availability as the number of replicas increases. Hence, in order to provide faster response time, most of the distributed storage systems rely on eventual consistency and do not satisfy the intersection property by default. In order to ensure strong consistency on demand, the system lets database users choose the number of nodes to be contacted during read and/or write operations.

The existing "strong" consistency options offered by these storage systems are strong enough to ensure data consistency when there is no network partition, at the expense of some extra communication cost and a risk to system availability. In these type of systems, there is no intermediate option that can ensure a better level of consistency with availability and latency guarantees similar to those of eventual consistency.

To overcome this limitation, the LibRe protocol was designed to act as an intermediate consistency strategy between the default eventual consistency and the strong consistency options derived from the quorum intersection property. In this paper, we extend our previous works on the LibRe protocol (Kumar et al. 2013, 2015) by providing a formal proof that the protocol is reliable in the face of nodes failures. In addition to these theoretical results, we provide a simulation-based scalability evaluation of the protocol, up to one hundred nodes. We show that, since LibRe is designed to provide higher availability and minimum request latency, its consistency guarantees are slightly relaxed when compared to other protocols. The following section describes the state of the art of consistency protocols that are based on quorum voting principle. The LibRe protocol is described in Sect. 3, followed by a detailed formal description in Sect. 4. Section 5 provides a scalability evaluation of the protocol through simulation, and the implementation of the LibRe protocol inside Cassandra distributed data storage system and its performance evaluations are discussed in Sect. 6 leaving the conclusion remarks in the end.

2 State of the Art

In quorum-based voting systems, consistency is ensured based on the size (number of members) of the read and write quorums. The size can be specified explicitly by a certain number such as 1, 2, 3 or can be left implicit by using a general term *all*, *quorum* (more than half). The node responsible for querying appropriate replica nodes and forwarding the result back to the client is known as the *coordinator node*. In this section, we describe some of the popular consistency protocols based on quorum-based voting principle.

- **Majority Quorum**: In majority quorum systems, the right replica version is chosen based on majority voting (Thomas 1979) for both read and write operations. Since both read and write operations are acknowledged by a majority of the replica nodes (quorum), the system ensures the intersection property and can guarantee consistency.
- **Weighted Voting**: Unlike majority quorum systems that assign equal vote (usually one) to each replica node, weighted voting systems (Gifford 1979) assign a variable number of votes to each replica node. The system ensures consistency based on the formula $R + W > V$, where R denotes the number of votes obtained during read operation, W denotes the votes gathered during write operation and V denotes the total number of votes assigned to the particular data item. By assigning lower number of votes to suspicious nodes that tend to fail often or get partitioned, the system can tolerate a higher number of node failures and still guarantee the intersection property.
- **ROWA**: ROWA stands for Read-One, Write-All protocol (Helal et al. 2002; Bernstein et al. 1987). As the name suggests, the read operations will be accomplished on only one of the replica nodes, whereas write operations have to be accomplished on all the replica nodes. The protocol is very efficient for read-heavy workloads. Since the read operations have to contact only one of the replica nodes, clients can choose to connect to the closest replica node to reduce the latency of read operations. The main limitation of this model is the additional overhead incurred during write operations.

- **ROWA-A**: In order to address the availability limitation of the write operations in ROWA protocol, ROWA-A was proposed in (Helal et al. 2002; Burckhardt 2014). ROWA-A stands for Read-One, Write-All Available. Unlike ROWA, that rejects the write operation if one of the replica nodes fails to acknowledge, a write in ROWA-A succeeds if at least one of the replica nodes acknowledges the operation. The coordinator node then records a *hint* for each replica node which failed to acknowledge the write operation. When the replica node that missed a write operation joins back the cluster, the coordinator node *hands-off* the missed write operation using the registered hints. This process is popularly known as *Hinted Hand-off* (Wiki 2013). The main limitation of this protocol is that it requires a reconciliation mechanism in order to correctly propagate the modifications when partitioned nodes are joining back the cluster.
- **Missing Writes Protocol**: In order to overcome the limitations that occur during network partitions with the ROWA-A protocol, the missing writes protocol (Eager and Sevcik 1983; Helal et al. 2002) was designed, combining both ROWA and majority quorum approaches. According to the missing writes protocol, the system follows the ROWA strategy when all nodes are available. When a failure is detected, write operations are applied only if a majority of nodes is available. Therefore, in case of partition, only the client that can reach a majority of nodes will be able to apply updates.
- **Epoch Protocol**: One of the limitations of the ROWA and ROWA-A protocols is that write operations have to be accomplished on every replica node, affecting the latency and availability of the write operations. Although quorum-based protocols minimize this impact by contacting only a majority of the replica nodes, the number of messages exchanged still depends on the total number of replicas. Hence, in order to address these limitations, a new approach for replica control called epoch protocol (Rabinovich and Lazowska 1993) was designed. In the epoch protocol, some nodes that are trusted to be operational at a previous period of time, are called *epoch members*. During write operations, instead of communicating the operation to all or a majority of replica nodes, it is only sent to the current epoch members (which are a relatively small set of replica nodes). Read operations are sent to members of the current epoch set, ensuring the read of the last write operation. However, the system has to run an epoch verification process periodically in order to check failures of epoch members. If one of the epoch members has failed, the protocol forms a new epoch that includes at least a majority of nodes from the previous epoch.
- **Probabilistic Quorum**: The idea of probabilistic quorum protocol (Malkhi et al. 1997) is to relax the need for quorum intersection for some requests based on probabilistic evaluation. In other words, probabilistic quorum protocols try to satisfy the intersection property only if the probability that a particular request leads to inconsistency is higher than a critical threshold. Some approaches exist in the literature to evaluate the probability that a particular request leads to inconsistency. For example, in (Bailis et al. 2012), Bailis et al. evaluate probabilistic bounded staleness based on the time taken for a particular write operation to be propagated to all the replica nodes. If a read operation for the last written data item arrives before

this probabilistic propagation time, then the read would be probably stale. Hence, based on the statistics about read/write pattern of the application or data-store, it is possible to decide whether a particular request requires voting or not.

- **Partial Quorum**: Partial quorum systems (Bailis et al. 2012) contact only one of the replica nodes for both read and write operations in order to favor high availability and minimum request latency during read and write operations. In this model, the system ensures eventual consistency, ensuring convergence with temporary inconsistencies. In order to ensure strong consistency on demand, the user/application can use Consistency-option ALL either during read or write operation to satisfy the formula $R + W > N$ (Intersection property). For a *read-heavy workload*, choosing consistency option ALL for writes and consistency option ONE for reads would be beneficial. For a *write-heavy workload*, choosing consistency option ALL for reads and consistency option ONE for writes would be beneficial.

In all these works, in order to ensure data consistency, a majority of nodes has to be contacted during read and/or write operations, thus impacting the request latency. The LibRe protocol studied in this paper provides a new approach to ensure data consistency while contacting only a minimum number of replica nodes during read and write operations, with the help of a registry look-up.

3   The LibRe Approach

The LibRe acronym stands for "Library for Replication". The goal of this approach is to devise a consistency protocol providing better latency in exchange of a slight decrease in consistency guarantees. The "library" part in "LibRe" refers to the use of a *registry* in the system. This *registry* stores, for each data item in the system, an *identifier*, a *version-id* and a set of nodes. The *version-id* is a monotonically increasing value representing the most recent version of the data item. For example, it can be the time-stamp of the operation or a version number. The set of nodes contains the addresses of the nodes that store the data item with the version indicated by the *version-id* in the registry entry. This registry is updated by the nodes every time they update a data item stored locally. Then, during read operations, looking up the registry enables to find quickly which replica holds the most recent version of the target data item. This section provides a detailed description of the LibRe protocol working principle.

3.1   Targeted System

Key-Value data stores have received increasing attention in modern distributed storage systems. Some of these systems ensure strong consistency, but most of the popular systems such as Dynamo (DeCandia et al. 2007), Cassandra (Lakshman and

Malik 2010), Riak (Klophaus 2010) and Voldemort (Voldemort 2015) rely on configurable consistency. These systems allow either the administrator or the developer to configure the required level of consistency for a given query or table.

Clients access the system through a *Coordinator Node*, this particular node is simply a point of entry, or front-end, to the system. The role of the *Coordinator Node* is to forward the request to the appropriate *Replica Nodes* depending on the data key.

In order to tune the consistency level, these storage systems follow a quorum-based voting principle such as described in Sects. 1 and 2. The level of consistency is thus configured by giving the minimum number of nodes required to acknowledge a query. The other characteristic of these systems is the reliance on a Distributed Hash Table (DHT) (Zhang et al. 2013). This allows any node in the system to locate a data item, provided it knows its key. LibRe assumes the presence of fault detection mechanisms providing knowledge about the state of the system (cluster membership), and reconciliation mechanisms that synchronize data items when the system recovers from a partition.

3.2   LibRe Registry

The core of the LibRe protocol is its "Registry". The registry is an in-memory data structure that maps a data identifier to the set of replica nodes identifiers holding the most recent *version id* of the data item. During a write operation, the *id* of each replica node that successfully wrote the data item is added to the set. If the size of the set reaches the total number of replica nodes in charge of that data item, confirming the convergence of all replicas, the entry for the data item in the registry is removed.

The registry is distributed over all the nodes in the cluster, but not replicated itself. Figure 1 shows the position of LibRe in the system architecture and the components

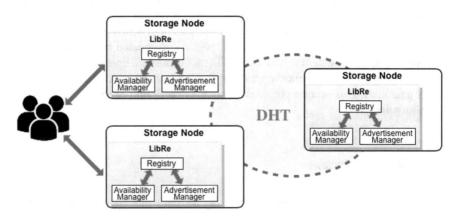

Fig. 1 LibRe architecture diagram

of the LibRe protocol. Given a data item d with key k, and a consistent hashing function h, let $h(k) = \{r_1^k, r_2^k, ..., r_n^k\}$ be the replica set (the set of nodes) in charge of d, where r_x^k denote node identifiers, and n is the number of nodes in the replica set. In each replica set, one of the nodes holds the registry and the other two supporting components of LibRe: the Availability-Manager and the Advertisement-Manager depicted in Fig. 1. This node is called the Registry Node for data d, and it is the first replica node that is obtained via $h(k)$. In the remainder of this document, it will be denoted as r_1^k. The LibRe protocol relies on the fact that, at any point in time, there exists only one copy of the registry entry for a data item. The registry is distributed over all nodes: each replica node holds a registry entry for data items where the node is the first replica node.

3.3 LibRe Messages

The LibRe protocol is based on two types of messages, namely: Advertisement Message (Fig. 2a) and Availability Message (Fig. 2b), corresponding respectively to the Advertisement-Manager and Availability Manager of the LibRe components shown in Fig. 1.

- **Advertisement Message**: As seen in Figs. 1 and 2a, clients connect to the Coordinator node, which may be different from the Registry Node. Usually, in a quorum based replication system, a write request is forwarded to all available replica nodes. If the Coordinator Node receives back the required number of acknowledgments (or votes) from the replica nodes for the write, the coordinator issues a success response to the client. If the sufficient number of vote is not received within a given time period, the coordinator issues a failure response to the client. LibRe protocol follows the same behavior, but in addition, after a successful write operation, each replica node sends an *Advertisement Message* to the Registry Node asynchronously, provided the data is configured to use LibRe. The advertisement message consists of the data key, version id and the originating node id.
- **Availability Message**: When the Coordinator Node receives a read request (Fig. 2b), that is configured to use LibRe protocol, it sends an *availability message* to the Registry Node of this particular data item. The availability message contains the original *read message* received from the client and the data item's key. When the Registry Node receives an *Availability Message*, it finds a replica node from the registry using the data key and forwards the original *read message* to that node. The replica node sends the read response directly to the Coordinator Node, which forwards it to the client. If an entry for a data key is not found in the registry, then the *read message* is forwarded to one of the available replica nodes.

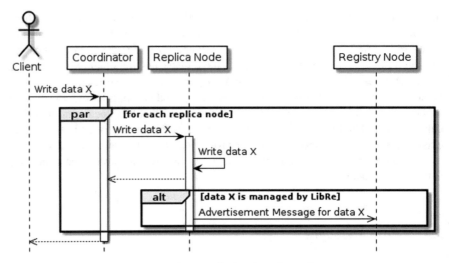

(a) LibRe Write/Update Operation

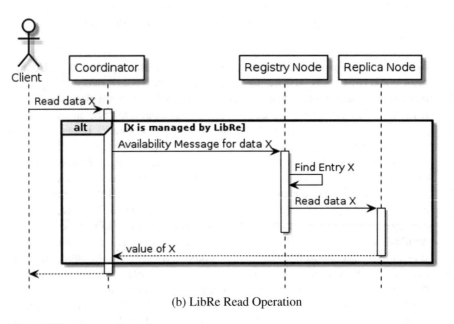

(b) LibRe Read Operation

Fig. 2 LibRe sequence diagram

3.4 LibRe Write Operation

Algorithm 1 describes the behavior of LibRe's *Advertisement Manager* that is triggered every time the Registry Node receives an Advertisement Message during an update operation. Note that an update for a data item can be issued (by the Coordinator) when the replicas are either in converged state or in diverged state.

Algorithm 1: Update Operation (*Registry Node*)

Input: Segment: (k, v, n), the key and version id of the data item and the sender's node id.
$R : k \rightarrow e_k$: the Registry. It maps keys k to corresponding entry record e_k,
$e_k = \langle v_k, r_k \rangle$: record, where v is a version-id and r_k is a replica set,
$r_k = \{n_1, n_2, \ldots\}$: set of nodes holding the given version of data item k.
N: Number of replicas

1 **Function** advertisementManager (k, v, n):
2 | *entry* $\leftarrow R.k$
3 | **if** *entry* $\neq \emptyset$ *and entry.v* $= v$ **then**
4 | | *entry.r* $\leftarrow$ *entry.r* $\cup \{n\}$
5 | | **if** $|entry.r| = N$ **then**
6 | | | *entry* $= \emptyset$
7 | **else if** *entry* $= \emptyset$ *or* $v >$ *entry.v* **then**
8 | | *entry.v* $\leftarrow v$
9 | | *entry.r* $\leftarrow \{n\}$
10 | $R.k \leftarrow$ *entry*

When a replica node n sends an Advertisement Message, the *Advertisement Manager* of the Registry Node takes on the following actions. First, the protocol checks whether the *data key* k is present in the *Registry* (line 3). If the key exists in the registry, it implies that the replicas are in diverged state. Then checks whether the *version id* in the registry entry matches the *version id* received with the message (line 3). If this is the case (replica convergence), then, line 4, the *node id* (IP-address) is appended to the current replica set for k. Next, lines 5–6, if the number of entries in the set reaches the total number of replicas, we are in a converged state for data key k and the entry for this key is deleted.

Line 8, if the entry does not exist in the registry or the *version id* received is greater than the *version id* in the entry, a new entry is created and initialized with the operation's *version id* and the sender's *node id*, lines 9 and 10. Finally the registry is updated, line 12. Depending on the nature of the *version id*, this algorithm achieves Last Writer Wins policy (Saito and Shapiro 2005).

3.5 LibRe Read Operation

Algorithm 2 describes the LibRe policy during a read operation. Since the Registry stores the up-to-date replica nodes' id, the protocol begins by fetching the registry entry for data item k, line 2. Line 3, if the entry for k exists, then, line 4, one of the replica nodes from the entry is chosen as the target node. The method *closest()*, lines

4 and 6, returns the closest replica node based on proximity. Line 5, if there is no entry for k, then line 6, one of the replica nodes that are responsible for storing the data item will be retrieved locally via DHT look-up, assuming every node is up-to-date. Finally, line 8, the read message is forwarded to the chosen target node.

Algorithm 2: Read Operation (*Registry Node*)

Input: (k, M_{read}), the key and the original read message.
$R : k \rightarrow e_k$: the Registry (see Algorithm 1).
$D_k = \{n_1, n_2, ...\}$: replica nodes for data item k obtained via the default method.

1 **Function** getTargetNode(k, M_{read}):
2      $replicaNodes \leftarrow R.k.r$
3   **if** $R.k \neq \emptyset$ *and* $replicaNodes \neq \emptyset$ **then**
4          $n \leftarrow replicaNodes.closest()$
5   **else**
6          $n \leftarrow D_k.closest()$
7          $n \leftarrow getCloser_Default(k)$
8   forward(M_{read}, n)

3.6 LibRe Reliability

As mentioned in Sect. 3.1, the LibRe protocol relies on the underlying database system to get knowledge about faults. In the case of DHT-based key-value stores, the system actively monitors ring membership and node failures (Hayashibara et al. 2004). In the event of a node joining and/or leaving the cluster, the DHT mechanism ensures a minimal redistribution of data keys over the ring members. In such case, there will be a change in the first available replica node (registry node) for a few data items, and consequently the registry information for these data-keys would not be available. In that case, the remaining node in the replica set with the lowest id is designated as the new registry node. However, this new registry node does not have the latest update information, thus the corresponding entries will be created by the subsequent *advertisement messages* sent to this node. This may lead to small and time-limited inconsistencies in the system. Therefore, LibRe sacrifices consistency in favor of availability: cf. Algorithm 2. If a registry node that has been unavailable joins back the cluster, the stale registry information is emptied. Besides, periodic local garbage collection has to be performed to keep the registry information clean between replica nodes.

3.7 LibRe Cost

The trade-off provided by the LibRe protocol comes at the expense of some additional message transfers and memory consumption.

3.7.1 Extra Message Transfers

In LibRe, a look-up in the registry is required during a read operation on contacting the *availability manager* of the registry node to read from a right replica node. However, this operation represent constant cost, as the number of messages sent for achieving the consistent read does not depend on the number of replicas involved, as mentioned in Sect. 3.3. During write operations, notifying the *advertisement manager* about an update is asynchronous and thus does not affect the write latency. Although these messages are an additional effort when compared to the default behavior of the system, it is better than the strong consistency options that communicate with a majority or all replica nodes during reads and/or writes.

3.7.2 Registry In-Memory Data Structure

LibRe manages the registry information in-memory. This information is distributed among all nodes in the cluster and is maintained only for the data items that are not yet up to date on all replicas. Moreover, eventual consistency guarantees of the targeted system and the periodic local garbage collection of the LibRe protocol help to reduce the amount of information to be kept in-memory.

4 LibRe Formalization

In this section, we formalize the LibRe protocol and provide a proof of its reliability. This work requires the introduction of notations that will be used in the remainder of this section.

4.1 System Model and Notations

A data item is a pair of key and value $\langle k, v \rangle$, where $\mathcal{K}$ is the set of keys. The value of a data item k depends on time, so we must add this parameter to our model. Time can be physical or logical (version number), the only property needed on the set of times $\mathcal{T}$ is total ordering. We therefore introduce a *value function*, indexed by keys and time, whose range is a set of values $\mathcal{V}$:

$$value : \mathcal{K} \times \mathcal{T} \rightarrow \mathcal{V}$$
$$(k, t) \mapsto v$$

We make the assumption, throughout this section, that *all written values are distinct*, this can be obtained by tagging values themselves with the time of the write request, so considering $\mathcal{T} \times \mathcal{V}$ instead of $\mathcal{V}$ as the set of values.

Thus, a data item *at time t* is the pair key/value $\langle k, value(k, t) \rangle$, denoted as k_t, which is the *latest version* of k at time t. We consider a *domain* $\mathcal{D} = \{k^1, \ldots k^m\}$ of items (represented by their keys) managed by LibRe. We also define the *latest modification time* with respect to time t, as:

Definition 1 Given a key k and a time t, we let $latest(k, t) = \min\{t' \in \mathcal{T} \mid k_{t'} = k_t\}$.

In particular, for any $t'' \in \mathcal{T}$, if $latest(k_t) \le t'' \le t$, we have $k_t = k_{t''}$. This includes $t' = latest(k_t)$ itself and this is the minimal such time, for which $k_{t'}$ is still the latest version of k (at time t). This is directly related to the last write request and represents an idealized model of storage. We will also denote $latest(k, t)$ the current data itself, $\langle k, value(k, t) \rangle$, which is the latest version of the data, that the system should provide when requested by the user. We can also compute a *version number*:

Definition 2 Given a key k and a time t, we let $version(k, t)$ be

---

```
1 Function version (k, t):
2     if t = 0 then
3         return 0
4     else
5         t' ← latest(k, t)
6         return (1+ version (k, t' − 1))
```

As written values are distinct, we can define the write time of a value $v \in \mathcal{V}$:

Definition 3 Given a key k and a value v, let

$$time(k, v) = \min\{t \in \mathcal{T} \mid value(k, t) = v\}$$

t is uniquely determined, and we have in particular $time(k, v) = latest$ $(k, time(k.v))$. It can be undefined in case v was never associated to k. This also implies a new definition of the version number:

Definition 4 Given a key k and a value v, let $version(k, v) = version(k, time$ $(k, v))$.

A replica node, denoted r, is a partial map from key to values, that depends on time. So its type is:

$$\mathcal{T} \to (\mathcal{K} \hookrightarrow \mathcal{V})$$

The state of the map at time t is $r(t)$, also denoted r_t, and dependency on time is usually left implicit. It is a partial mapping (notation $\mathcal{K} \hookrightarrow \mathcal{V}$) since it is undefined on keys it is not in charge of. A read operation for k on that map is then $r_t(k)$ and a put operation is denoted $r_t[k \leftarrow v]$. It has the effect that r_{t+1} is the same map as r_t, except on k.

Remind, from Sect. 3.2, that the set of replica nodes in charge of an item k is given by applying the consistent hashing function $h(k) = \{r_1^k, ..., r_n^k\}$. So the Registry Node for k is r_1^k, more compactly noted r_k, when no confusion arise.

The registry of a node r is called $\rho = reg(r)$. ρ bears the same indices as r, in particular, the registry responsible for k is $\rho_k = \rho_1^k = reg(r_1^k)$. Registries also vary over time, therefore are function of the following type:

$$\mathscr{T} \to (\mathscr{K} \hookrightarrow \wp(\mathscr{R}))$$

$\rho(t)$, as usual noted ρ_t or ρ when the dependency on time is implicit, is a *partial* function. It associates a subset of $\mathscr{R}$ (the nodes) to *some* keys, at most the ones the node r is in charge of. As discussed in Sects. 3.4 and 3.5, even $\rho_k(k)$ may be undefined, or more generally $\rho(k')$ for any arbitrary k' the registry ρ is in charge of. The intended meaning is that, in such a case, the Registry Node *knows* that the replicas for k' are in a consistent state. We prove this claim later in the Section.

We also need to introduce the function R_l, that to each data item k associates the *actual* set of replica nodes that contain the latest version of k (at a given time t):

$$R_l(k, t) \subseteq h(k),$$
$$r \in R_l(k, t) \text{ if and only if } r_t(k) = value(k, t)$$

As we are assuming that all written values are distinct, it cannot be the case that $R_l(k, t)$ contains some stale node r, for which $r_t(k) = value(k, t')$ for $t' < latest(k, t)$, but still enjoying $value(k, t') = value(k, t)$.

By *definition* of $R_l(k, t)$, the following holds: $time(k, r_t(k)) < t$, for any $r \in h(k)\backslash R_l(k, t)$, i.e. $r \in h(k)$ but $r \notin R_l(k, t)$. Otherwise, r would belong to $R_l(k, t)$. Otherwise said, those r not in $R_l(k, t)$ contain stale data for key k, for which $version(k, r_t(k)) < version(k, t)$.

The operations are of two kind, get (g) or put (p). They arrive at various time, as a discrete stream. Each operation contains: a time t; the data key k; and put operations also contains the new value v. We access those value with a dot: $p.t, p.k$ and $p.v$. All put operations p enjoy the following property:

$$value(p.k, p.t) = p.v$$

We also assume, for any key k and time t (note that $p.t$ is different from t):

if $latest(k, t) \neq 0$ then $\exists p, p.k = k$ and $p.t = latest(k, t)$ and $p.v = value(k, t)$

Lastly, we assume *no interference* of a put operation p over the state of the system (registries, data on replica) associated with a key $k' \neq p.k$.

4.2   System States

The system can be in one of the following two states with the associated properties.

4.2.1   Stable State: S1

The system is said to be in stable state at t if two conditions are met. The first one is that there is *no change* in ring membership compared to $t - 1$, and the second one is that the system enjoys (at t) for any k the two following properties:

$$\rho_k(k) \text{ defined } \Rightarrow \emptyset \neq \rho_k(k) \subseteq R_l(k) \tag{1}$$

$$\rho_k(k) \text{ undefined } \Rightarrow h(k) = R_l(k) \tag{2}$$

As those Properties 1 and 2 are mutually exclusive, we will ensure only one of them at a time for a given k. In case either property holds for k, we say that the registry for k is *up to date*, or simply that the registries are up to date, if this holds for all $k \in \mathcal{K}$. Note that in Property 1, ρ_k can be *strictly smaller* than $R_l(k)$, in case there was a communication problem (cf. Lemma 1). Moreover, in state S1, we know that the last version of k is present on some replica node, and accessible to the other nodes. As a consequence:

$$R_l(k) \neq \emptyset \tag{3}$$

4.2.2   Unstable States: S2, S3, S4 and S5

We denote the state, when a node r *leaves unexpectedly the ring* or becomes inaccessible, at time t, by S2. In this case, there is a switch of the registry for the following set of keys:

$$sup(\rho_{t-1}) = \{k \text{ such that } h_{t-1}(k) = \{r, r_2, \ldots, r_n\}\}$$

$sup(\rho)$ is the support of ρ, the set of all keys it can be defined on. It is larger than $dom(\rho)$, the *domain* of ρ, which is the set of keys on which ρ is currently defined. An undefined $\rho_{t-1}(k)$ for $k \in sup(\rho)\backslash dom(\rho)$ just means that $R_l(k, t - 1) = h(k)$ (replicas have converged, cf. Property 2). For such a key k, removing r has virtually no impact. Generally speaking, we get, for $k \in dom(\rho)$:

$$\rho_k(t)(k) \neq R_l(k, t)$$

In more details, fix some key k such that $h_{t-1}(k) = \{r, r_2, \ldots, r_n\}$. As r has left the mode, $h_t(k) = \{r_2, \ldots, r_n\}$, and $\rho_k(t) = \rho_2(t)$: the registry node has changed, as specified in Sect. 3.6.

Thus, when we are in state S2, Algorithm 2 will pick any of the nodes of $\rho_2(t)(k)$ or $h(k)$, with a risk of staleness with probability that depends on $|R_l(k)|/|h(k)|$, see Sect. 4.4.

However, in state S2, for any k, in particular those in $dom(\rho)$ the following property holds, which is formally proved in Lemma 1.

$$\forall p, (p.t > t \wedge p.k = k) \Rightarrow \forall t' > p.t + \tau_{adv}, R_l(k, t) = \rho_2(t)(k) \qquad (4)$$

This means: as soon as we have a put operation on k, at any time after at least one advertisement message has been received (hence the delay τ_{adv} added to the time $p.t$ the request was made), the registry ρ becomes up to date for k (provided other errors do not occur in the frame $[t, p.t + \tau_{adv}]$).

We denote the state when a node joins the ring by S3. In this case, there is also a switch on the registry for certain keys k. The system *can* fall back in the state $\rho_k(t)(k) \neq R_l(k, t)$, especially if the new node joins the ring without synchronization. The system enjoys the same Property 4.

The last unstable states, S4 and S5 are similar: the Registry Node r receives no advertisement message and the Coordinator Node c receives the required number of acknowledgments (S4), or r receives at least one advertisement and c receives not enough acknowledgments (S5). Those states are unlikely to happen, as it is rather an indicator that nodes are inaccessible (state S2). All states S2 through S5 produce same effect on the system, that is to say:

$$S2 \approx S3 \approx S4 \approx S5$$

There are other situations, when a write operation fails. One last instance is when no replica is available, in which case the Coordinator will issue a failure, so this does not produce any consistency issue.

As for the likeliness of state S4, a put operation is sent to all the replica nodes in parallel (including r_k), and the local advertisement message is likely not to be lost. In the case where the put operation sent to r_k is lost, the registry ρ_k will be updated on reception of at least one of the advertisements sent by the replica nodes that have successfully voted to the Coordinator (assuming that if the Coordinator receives the acknowledgment, the Registry receives the advertisement). In the worst case, the put operation sent to r_k is lost and all the advertisement message sent to r_k by the replica nodes, that succeeded, are lost as well, and we fall in state S4.

As for state S5, such a situation might arise more frequently. This is less harmful, since the nodes pointed to by the Registry Node for k would contain newer data, than the one given by the last successful write.

4.3   Stable State Properties

We assume to be on a time range $[t_1, t_2]$, such that there is no change in the ring membership during the time frame $[t_1, t_2]$. As well, we assume that all the put operations received in $[t_1, t_2]$ are successfully advertised at least once to the Registry Node *and* that the Coordinator issues a success response to p. Otherwise, we fall in one of the unstable states, discussed in the next section.

Given a put operation p, we define τ_{adv}, abbreviated as τ, the delay with which the *first* advertisement message is received by the Registry Node. As assumed there is at least one such message.

We show that the users always read the most recent version of any data item k, except during the time when a put operation is in process on k. We begin with lemmas *that do not depend on the state*, and that we will reuse later. The next lemma does depend only on the success of p (and on no change on ring membership):

Lemma 1 *Consider a put operation p. $\rho_{p.k}$ is up to date for $p.k$ at time $t + \tau_{adv}$, provided that we do not receive another put operation p' on k in $[p.t, p.t + \tau_{adv}]$.*

For simplicity, we assume no concurrent put operation p' in the interval between the reception of p and the first advertisement is successfully sent to $r_{p.k}$. In practice, LibRe takes care of overlapping put operations by applying the Last Writer Wins (LWW) policy with the help of the tie breaking mechanism provided by the underlying system. To take this into account, we need only to consider a set of overlapping put operations, and ensure up-to-dateness wrt the last one.

Proof During the operation, the Coordinator sends the update in parallel to all the replica nodes in $h(p.k)$ and waits for sufficient number of votes (node acknowledgments for successful write) according to the chosen consistency option. A success message is issued to the writer as soon as the threshold is met (at time $t' > p.t$).

We assume that $h(k)$ contains more than one replica. By Algorithm 1, $\rho_{p.k}(p.k)$ is set to $\{r\}$, where r is the source of the first advertisement message, received at $p.t + \tau$. Therefore $\rho_{p.k}(p.k)$ is not empty, and by definition: $r \in \rho_{p.k}(p.k)$ implies data $\langle p.k, p.v \rangle$ has been written.

Since there is no overlapping put operation on $p.k$, $p.v = value(p.k, p.t) = value(p.k, p.t + \tau)$, therefore $r \in R_l(p.k, p.t + \tau)$ and $\rho_{p.k}(t + \tau) \subseteq R_l(p.k, t + \tau)$. Property 1 holds for k.

Lastly, if $h(k)$ contains only one replica, then it must be r, the advertiser. We know, by the same reasons as above, that $r \in R_l(p.k, p.t + \tau)$, as a consequence $R_l(p.k, p.t + \tau) = h(k)$. $\rho_{p.k}(t + \tau) = \emptyset$ by the algorithm, which means, that Property 2 is respected.

This covers all the possible case that can happen during the execution of a successful put operation, including loss of put messages sent to the $r \in h(p.k)$ (including $r_{p.k}$ itself), of advertisement messages sent to $r_{p.k}$, and of acknowledgment messages sent to the Coordinator. The important point is that the Registry Node is up to date as soon as one advertisement message is received.   □

The next lemma says that up-to-dateness is hereditary in time. It does not depend on the success of any put operation, nor on the state we are in, and only assume stability of ring membership:

Lemma 2 *Let k be a key, assume that ρ_k is up to date for k at t. Let $t' \geq t$. If no put operation on k is received between t and t', then ρ_k is up to date for k at t'.*

Proof This proof is by induction on t':

- if $t' = t$, then ρ_k is up to date (for k, left implicit from now on) at t' by assumption.
- assume that ρ_k is up to date at t', let us show that it is up to date at time $t' + 1$. Several cases must be distinguished:

  - $\rho_k(t')(k) = \emptyset$. Then all the nodes of $h(k)$ contain the last version of k, by Property 2. As no other put operation on k is received, $R_l(k, t' + 1) = h(k)$.
  - $\rho_k(t')(k) \neq \emptyset$. Then we distinguish the following cases:

    we receive no advertisement at $t' + 1$, or a stale advertisement: $\rho_k(t' + 1)(k) = \rho_k(t')(k)$ is still up to date.

    we receive a valid advertisement from a node r. ρ_k will be updated. As $\rho_k(t')$ was up to date, we have by Property 1, $\rho_k(t')(k) \subseteq R_l(k, t')$. We also know that, at $t' + 1$, r has successfully written the data, so $r \in R_l(k, t' + 1)$. Lastly, $R_l(k, t') \subseteq R_l(k, t' + 1)$. This implies $\rho_k(t')(k) \cup \{r\} \subseteq R_l(k, t') \cup \{r\} \subseteq R_l(k, t' + 1)$.

    We now distinguish two possibilities. First, if $\rho_k(t')(k) \cup \{r\} = h(k)$, we set $\rho_k(t' + 1)(k) = \emptyset$. But this also implies, by the results just proved, that $h(k) \subseteq R_l(k, t' + 1)$. Thus, $R_l(k, t' + 1) = h(k)$ (it cannot be larger) and Property 2 is respected. Otherwise, $\rho_k(t' + 1)(k) \subseteq R_l(k, t' + 1)$ and Property 1 is satisfied. □

This allows us to derive preservation of stability of a system on the interval $[t_1, t_2]$.

Lemma 3 *Assume that the system is in stable state S1. If we receive only successful put operations, then the registries are up to date at any time $t \in [t_1, t_2]$, if no put operation is ongoing at t. In other words, the system is in stable state S1, except when a put operation is ongoing.*

Proof Fix some key $k \in \mathcal{K}$. We first prove by induction on $t \in [t_1, t_2]$, that the registry for k is up to date at t:

- if $t = t_1$, the claim holds by assumption.
- if $t = p.t + \tau$ for some put operation on k, then the registry is up to date for k at t by Lemma 1.
- otherwise, we apply induction hypothesis, which gives us an up-to-date registry at time $t - 1$, and we apply Lemma 2. No operation was ongoing on k at $t - 1$, otherwise we would be in the previous case, so we can apply the induction hypothesis.

This holds for any k, as soon as no put on k is ongoing at t. Therefore, if no operation is ongoing overall, this holds for any k, and the Lemma holds. □

Lemma 4 *Let k be a key and t be a time. n stable state, if no put operation is ongoing on k, ρ_k always returns a node id that contains $latest(k, t)$, the latest data.*

Proof According to the system properties under stable state (S1), there are two cases with respect to the Properties 1 and 2:

1. $\rho_k(k)$ is defined. As it is up to date, it returns some $r \in R_l(k)$.
2. if the entry for k *does not exist* in ρ_k, then ρ_k returns the id of any $r \in h(k)$. According to Property 2, $\rho_k(k)$ undefined implies $R_l(k) = h(k)$. ρ_k returns any $r \in R_l(k) = h(k)$, but all the replica nodes contain the latest version.   □

Theorem 1 *Let k be a key. If there is no ongoing write operation on k, then users read the most recent version of k.*

Proof From Lemma 4, ρ_k always returns a node id (IP address) that contains $latest(k, t)$, when in stable state. According to Lemma 3, the system is in stable state, except temporarily, when a write operation is ongoing. So users read the most recent version of k.   □

4.4 Unstable States Properties

When in an unstable state, information returned by the registry can be wrong in two ways:

- The registry does not contain a subset of $R_l(k)$, which means ρ_k is not up to date for k:

$$\rho_k \nsubseteq R_l(k)$$

  The probability that the registry returns an id (IP address) of a node that does *not* contains $latest(k, t)$ is $1 - \dfrac{|common|}{|\rho_k(k)|}$, where $common = \rho_k(k) \cap R_l(k)$.
- The registry contains no entry for k. The probability that the registry returns the id of a node that does *not* contains $latest(k, t)$ is $1 - \dfrac{|R_l(k)|}{|h(k)|}$.

Lemma 5 *Let r be a node with registry ρ, and $k \in dom(\rho)$. When r leaves the ring, the registry entry $\rho(k)$ will be rebuilt on the new registry node ρ_k after a successful write operation on k.*

Proof In the event of node joining or leaving the cluster (including nodes failure/disconnection), the underlying system guarantees helps to detect it at the earliest. In such a case, there will be a change in the first available replica node as mentioned in Sect. 4.2.2. As soon as this change has been made, and assuming that no node joins or leaves the ring, we are in the conditions of Lemma 1, that guarantees that the registry ρ_k is up to date for k after a write operation, and this property is hereditary

(Lemma 2), so up-to-dateness for k holds back (see the proof of Lemma 3), as long as all put operations on k are successful, and there is no new change in the ring membership. ☐

In particular we might encounter some $g(k, t)$, with $t \geq t_0$ (t_0 is the failure time of r), for which ρ_k returns $r \notin R_l(k)$. However, as we just saw, after the first $p(k, v, t')$, such that $t' > t_0$, $\rho_k(k)$ is up to date again and preserves this property. Therefore, for all $g(k, t'')$ with $t'' \geq t' + \tau$, ρ_k returns $r \in R_l(k) \subseteq h(k)$.

Since the put operation following the stale read updates the registry, for all subsequent get operation, ρ_k returns an id (IP address) of a node that contains $latest(k, t)$.

Now, assume that r leaves (or joins with a stale registry, or gets a stale registry due to a write failure, for the purpose) the ring, with registry ρ, at t_0. Assume also, that a write operation happens on all the data $k \in sup(\rho)$ (the keys, of which node r had the charge), for which the new ρ_k was *not* up to date. The registry entry $\rho_k(k)$ will be rebuilt on the new registry node r_k and is now up to date. For the data $k \in sup(\rho)$, such that ρ_k (the new registry) was up to date from the inception (see discussion in Sect. 4.2.2), there is no need for rebuilding, so no need for a write operation. As a consequence, all registries are up to date and the system has switched back to the stable state $S1$. This yields the following theorem.

Theorem 2 *Assuming no further joining/leaving node event or write problem, the time taken by the LibRe protocol to recover from an unstable state is the minimum among the time for the next successful write operations on all stale data T_{st} and the time taken to resolve the staleness via underlying system's eventual consistency guarantees P_{st}.*
From any unstable state, we converge towards stable state S1 in time $\min(T_{st}, P_{st})$.

Proof Let the time, that a quorum based replication system resolves staleness by itself via read-repair and generic anti-entropy protocols, be P_{st}.

From Lemma 5, the registry for the stale data is rebuilt in the new registry node after the next write operation, and we need one write operation per stale data (T_{st}).

Either of those solutions ensures that all the registries are up to date. From the above two propositions, the time taken by the LibRe protocol to recover from the unstable state is $\min(P_{st}, T_{st})$. ☐

We have shown different states of the LibRe protocol $S1$, $S2$, $S3$, $S4$, $S5$ and the properties of the protocol under each state. During normal condition, when there is no change in the ring membership and no write failure, the protocol remains in state $S1$. LibRe under state $S1$ ensures consistency of the read operations and we call it as *stable state*.

When a node joins or leaves the ring, change in the ring membership affects the consistency guarantees of the protocol. The event of a node leaving (unreachable to) the ring is denoted as state $S2$ and the event of node joining (or recovering from disconnection) the ring is denoted as state $S3$. Since the system exhibits the same property under both states $S2$ and $S3$, the two states are denoted commonly as *unstable state*, along with other states resulting from failure of a write operation. LibRe under unstable state ($S2$ or $S3$) do not ensure consistency of the read operations. We also formally showed that protocol under unstable state will switch back to the stable state $S1$ within short period of time.

5 Simulation-Based Evaluation

To evaluate the scalability of the LibRe protocol, we extended the Simizer [1] simulator to allow consistency evaluation (Lefebvre et al. 2014). This tool enables to simulate various configurations of networks and nodes in order to test quickly any distributed protocol and enabled us to test LibRe with an increasing number of nodes. The protocol implementation follows the architecture described in Sect. 3. In order to get simulation results closer to the real inner workings of the Cassandra (Lakshman and Malik 2010) storage system, during write operations, data is cached in memory following a Least Recently Used (LRU) policy. During read operations, request latency is computed based on the time needed to retrieve data from memory or disk, depending on whether the data is cached or not. If the target data item is available in the cache, the time to retrieve it will be considerably reduced. If the data item does not exist on the machine, the request will be considered as failed.

The simulation test-bed consists of two sets of nodes: the client nodes and the server nodes. Client nodes send a sequence of read and/or write requests (one request after another) to a target server node called *Coordinator*. The number of requests sent by each client is limited by the clients' lifetime, which is determined by a random law. Network latency is computed randomly following a normal distribution. All nodes, clients and servers are on the same network, and subject to the same latency variations.

Each test is run 30 times, for 25, 50, 75, and 100 nodes, and 1000 data items. This low number of item was used for creating contention on accesses to the item. Reads and write requests are in equal proportions, and latency is measured from the client issuing the request. Moreover, each data item is replicated three times.

Figure 3 shows the maximum 95th percentile latency of 30 simulations for read and write requests with LibRe protocol. The graph shows that the latency of the LibRe protocol is not affected by the total number of nodes in the cluster, thus validating the scalability of our approach. Because of the longer path taken by read request, the latency of reads is slightly higher than the latency of write requests. In practice it appears that the average response time for reads with LibRe is close to the average response time for quorum-based consistency, while providing faster write times. The difference increases further in presence of failures, because LibRe does not need to wait for several acknowledgments before providing an answer to the user.

---

[1] https://github.com/isep-rdi/simizer.

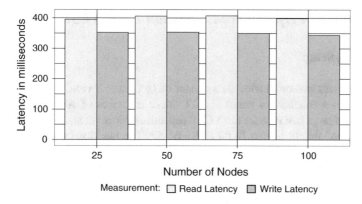

Fig. 3 Libre simulated read and write requests 95th percentile latency

Following these encouraging results, the LibRe protocol was implemented and tested in the Cassandra database system (Lakshman and Malik 2010).

6 CaLibRe: LibRe Implementation Inside Cassandra

Cassandra (Lakshman and Malik 2010) is one of the most popular open-source NoSql systems that satisfies the system model specified in Sect. 3.1. Hence, we decided to implement the LibRe protocol inside the Cassandra work-flow and evaluate its performance against Cassandra's native consistency options: ONE, QUORUM and ALL. Although Cassandra is a column family store, we used it as a Key-Value store during test setup detailed in Sect. 6.1.1. The LibRe protocol was implemented inside Cassandra release version 2.0.0. In the native work-flow, while querying a data, the endpoints (replica nodes) addresses are retrieved locally via matching the token number of the data item over the nodes token numbers. The IP-address of the first alive endpoint (without sorting the endpoints by proximity), will be chosen as the registry node for the replica sets it is responsible. LibRe messages are handled by separate thread pool. On system initialization, all LibRe registries are empty until a write request is executed, which will trigger the protocol and start filling up the registries. CaLibRe can be configured to work for specific data items, either by passing a list of data items in a configuration file, or by specifying directly the name of the table(s) on which the protocol should apply. Currently, the *version-id* used in CaLibRe is the hash of the written value. However, it will be replaced by the time-stamp in the future.

6.1   CaLibRe Performance Evaluation Using YCSB

6.1.1   Test Setup

The experiment was conducted on a cluster of 19 Cassandra and CaLibRe instances that includes 4 medium, 4 small and 11 micro instances of Amazon EC2[2] cloud service and 1 large instance for the YCSB test suite (Cooper et al. 2010). All instances were running Ubuntu Server 14.04 LTS - 64 bit. The workload pattern used for the test suite was the "Update-Heavy" workload (*workload-a*), with a record count of 100000, operation count of 100000, thread count of 10 and the Replication-Factor as 3. YCSB by default stores 10 columns per RowKey. We used RowKey as the data key, for which, an entry will be managed in the LibRe Registry. Using RowKey as the data key could leads to a situation like, if one or few columns of a RowKey is updated on a replica node r_n, the registry would assume r_n contains the recent version for all the columns of the RowKey. In order to avoid this situation, we update all the 10 columns during each update. The test case evaluate the performance and consistency of the 19 Cassandra instances with different consistency options (ONE, QUORUM and ALL) against 19 CaLibRe instances with a consistency option ONE. Performance is evaluated by measuring read and write latency and consistency is evaluated for each level by counting the number of stale reads. In order to simulate a significant number of stale reads, a partial update propagation mechanism was injected into the Cassandra and CaLibRe cluster to account for the system performance under this scenario (Kumar et al. 2014). Hence, during update operations, instead of propagating the update to all 3 replica nodes, the update will be propagated to only 2 of the replicas nodes.

6.1.2   Test Evaluation

Figure 4a, b and Table 1 respectively show the evaluation of Read Latency, Write Latency and the number of Stale Reads of Cassandra with different consistency options against CaLibRe: Cassandra with LibRe protocol. In Fig. 4a, the entity ONE represents the read and write operations with consistency option ONE. The read and write operations with consistency option QUORUM is indicated by the entity QUORUM. The entity ONE-ALL represents the operations with write consistency option ONE and read consistency option ALL. The entity CALIBRE represents our implementation of the LibRe protocol inside Cassandra. Due to the injection of the partial update propagation, ROWA (Read One, Write All) principle could not be tested, as writes will always fail.

In Fig. 4a, we can see that the 95th percentile read latency of CALIBRE is similar to the other consistency options of Cassandra. The 99th Percentile Latency of CALIBRE and Cassandra with consistency level ONE are lower than the other options ONE-ALL and QUORUM. The minimum and average latency of CALIBRE are slightly

---

[2]http://aws.amazon.com.

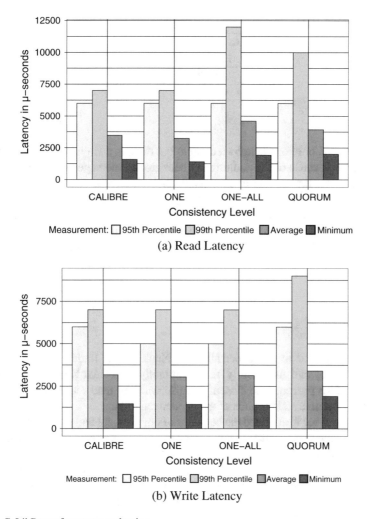

(a) Read Latency

(b) Write Latency

Fig. 4 CaLibRe performance evaluation

higher than Cassandra with consistency level ONE but better than the consistency
options QUORUM and ONE-ALL. This is due to the fact that LibRe protocol imposes
an additional call to the registry for all requests.

Figure 4b shows that the 95th percentile write latency of CALIBRE is the same as
the 95th percentile write latency of QUORUM, and that CALIBRE is faster in other
metrics: 99th Percentile, Minimum and Average latency of QUORUM. However,
while comparing to the entities ONE and ONE-ALL, it appears that some the write
latency of CALIBRE is slightly higher. This is due to the fact that both in ONE and
ONE-ALL, writes need only one acknowledgment from a replica node. In CALIBRE,

Table 1 Stale reads per
consistency option

Consistency option	Number of stale reads
ONE	405
ALL	8
QUORUM	9
CALIBRE	2

writes also need only one acknowledgment but there is a slightly increased load on replica nodes because of the additionally background messages sent to the registry.

Table 1 shows the number of stale reads for each level of consistency. Cassandra with consistency level ONE shows the highest number of stale reads. There were a few stale reads in the other consistency options, but these numbers are negligible when compared to the total number of requests. From these results, it is possible to conclude that CaLibRe offers a level of consistency similar to one provided by the QUORUM and ONE-ALL levels with better latency and availability.

7 Conclusion

The work described in this paper aims at enhancing the trade-offs between Consistency, Latency and Availability of an eventually consistent Key-Value store. Our protocol: LibRe prevents the system from forwarding read requests to the replica nodes that contain a stale replica of the needed data item.

In this paper, we formalized the behavior of our protocol and provided proofs that the system converges towards a stable and consistent state even when some nodes of the system fail or leave.

In order to assess the performance of the protocol, we evaluated its scalability by simulation from 25 to 100 nodes and showed that read and write latency do not degrade as the number of nodes in the system increases.

To evaluate further our protocol and compare it to other strategies, the 'CaLibRe' solution was implemented. This system offers the LibRe protocol as an additional consistency option for the Cassandra storage system. CaLibRe was evaluated against the native consistency options of Cassandra using YCSB on a 19 nodes CaLibRe and Cassandra cluster. The performance results prove that CaLibRe provides lower request latency compared to the strong consistency levels offered by Cassandra, combined to a similar number of stale reads. Hence, we can safely conclude that the LibRe protocol gives a new trade-off between consistency, latency and availability. However, the performance results were not evaluated under nodes joining or leaving the clusters. During such events, LibRe protocol would experience temporary inconsistencies as shown in Sect. 4.4.

This work provides insight on more subtle trade-offs between performance and consistency in distributed storage systems. The LibRe approach enables users to tune even more finely the storage system to the needs of the application. An interesting perspective to this work would be to study the influence of the nature of the version identifier (time-stamp, version vector, vector clocks, ...) on the stale read rate and the latency, in order to be able to tune even more finely the consistency level.

References

Abadi, D. J. (2012). Consistency tradeoffs in modern distributed database system design: Cap is only part of the story. *Computer, 45*(2), 37–42.

Bailis, P., Venkataraman, S., Franklin, M. J., Hellerstein, J. M., & Stoica, I. (2012). Probabilistically bounded staleness for practical partial quorums. *Proceedings of VLDB Endowment, 5*(8), 776–787.

Bernstein, P. A., Hadzilacos, V., & Goodman, N. (1987). *Concurrency control and recovery in database systems.* Inc, Boston, MA, USA: Addison-Wesley Longman Publishing Co.

Burckhardt, S. (2014). Principles of eventual consistency. *Foundations and Trends ® in Programming Languages, 1*(1–2), 1–150.

Cooper, B. F., Silberstein, A., Tam, E., Ramakrishnan, R., & Sears, R. (2010). Benchmarking cloud serving systems with ycsb. In *Proceedings of the 1st ACM Symposium on Cloud Computing, SoCC '10* (pp. 143–154). New York, NY, USA: ACM.

DeCandia, G., Hastorun, D., Jampani, M., Kakulapati, G., Lakshman, A., Pilchin, A., et al. (2007). Dynamo: amazon's highly available key-value store. *SIGOPS Operating System Review, 41*(6), 205–220.

Eager, D. L., & Sevcik, K. C. (1983). Achieving robustness in distributed database systems. *ACM Transactions on Database Systems, 8*(3), 354–381.

Gifford, D. K. (1979). Weighted voting for replicated data. In *Proceedings of the Seventh ACM Symposium on Operating Systems Principles, SOSP '79* (pp. 150–162). New York, NY, USA: ACM.

Gilbert, S., & Lynch, N. (2002). Brewer's conjecture and the feasibility of consistent, available, partition-tolerant web services. *SIGACT News, 33*(2), 51–59.

Hayashibara, N., Defago, X., Yared, R., & Katayama, T. (2004). The phi accrual failure detector. In *Proceedings of the 23rd IEEE International Symposium on, Reliable Distributed Systems, 2004* (pp. 66–78).

Helal, A., Heddaya, A., & Bhargava, B. K. (2002). Replication techniques in distributed systems. In *Advances in database systems* (Vol. 4). Kluwer.

Klophaus, R. (2010). Riak core: building distributed applications without shared state. In *ACM SIGPLAN commercial users of functional programming, CUFP '10* (pp. 14:1–14:1). New York, NY, USA: ACM.

Kumar, S., Lefebvre, S., Chiky, R., & Soudan, E. (2014). Evaluating consistency on the fly using ycsb. In *IWCIM, 2014,* (pp. 1–6).

Kumar, S. P., Chiky, R., Lefebvre, S., & Soudan, E. G. (2013). Libre: A consistency protocol for modern storage systems. In *Proceedings of the 6th ACM India Computing Convention, Compute '13*. ACM.

Kumar, S. P., Chiky, R., Lefebvre, S., & Soudan, E. G. (2015). Calibre: A better consistency-latency tradeoff for quorum based replication system. In *8th International Conference on Data Management in Cloud, Grid and P2P Systems*. Globe: Springer LNCS.

Lakshman, A., & Malik, P. (2010). Cassandra: A decentralized structured storage system. *SIGOPS Operating System Review, 44*(2), 35–40.

Lefebvre, S., Kumar, S. P., & Chiky, R. (2014). Simizer: Evaluating consistency trade offs through simulation. In *Proceedings of the First Workshop on Principles and Practice of Eventual Consistency, PaPEC '14*. New York, NY, USA: ACM.

Malkhi, D., Reiter, M., & Wright, R. (1997). *Probabilistic quorum systems, PODC '97* (pp. 267–273). ACM.

Naor, M., & Wool, A. (1998). The load, capacity, and availability of quorum systems. *SIAM Journal on Computing, 27*(2), 423–447.

Rabinovich, M. & Lazowska, E. D. (1993). An efficient and highly available read-one write-all protocol for replicated data management. In *Proceedings of the Second International Conference on Parallel and Distributed Information Systems, PDIS '93* (pp. 56–66). Los Alamitos, CA, USA: IEEE Computer Society Press.

Saito, Y., & Shapiro, M. (2005). Optimistic replication. *ACM Computing Surveys, 37*(1), 42–81.

Thomas, R. H. (1979). A majority consensus approach to concurrency control for multiple copy databases. *ACM Transactions on Database Systems, 4*(2), 180–209.

Voldemort, P. (2015). Physical architecture options. http://www.project-voldemort.com/voldemort/design.html.

Vukolic, M. (2010). Remarks: The origin of quorum systems. *Bulletin of the EATCS, 102,* 109–110.

Wiki, C. (2013). Hinted handoff. https://wiki.apache.org/cassandra/HintedHandoff.

Zhang, H., Wen, Y., Xie, H., & Yu, N. (2013). *Distributed hash table–theory, platforms and applications.* Springer Briefs in Computer Science: Springer.

Author Biographies

Dr. Sathiya Prabhu Kumar is a Research Engineer at Ecole Polytechnique, Palaiseau, near Paris, France, where he works on data science project in the Public Health System of France. His area of interest includes but not limited to Distributed Systems, Big Data, Database Systems and NoSql Systems. He holds a master degree in Software Engineering from VIT University, Vellore, India and a Doctoral degree in Computer Science from CNAM, Paris, France. His teaching activities include topics such as Advanced Databases, Distributed Systems, and Big Data tools.

Dr. Sylvain Lefebvre is associate professor in the RDI team of LISITE laboratory in Paris, France. He holds a master's degree in distributed systems engineering (PARIS XII, 2010), and Phd in computer science from CNAM, PARIS. His research works is focused on distributed load balancing, storage and data analysis systems.

Raja Chiky is currently Associate Professor at ISEP where she is head of the RDI team. She holds a Ph.D. in Computer Science from Telecom ParisTech obtained after a Master degree in data mining and an engineering degree in computer science. She worked closely with EDF R&D on research projects related to data stream mining. Her research interests include statistics, data mining, data warehousing, data stream management, recommender systems, and cloud computing

Dr. Olivier Hermant is a Researcher at MINES ParisTech. He is an expert in the field of formal methods, ranging from automated theorem proving to the foundations of programming languages and logical frameworks. He leads the development of the Dedukti type-checker and recently became involved in applying the results of his research to formalization of properties and to industrial problems. He holds a Ph.D. from Paris-Diderot University.

A Fault Detection Approach for Robotic Systems Combining the Data Obtained from Sensor Measurements and Linear Observer-Based Estimations

Rabih Taleb, Rabah Mazouzi, Lynda Seddiki, Cyril de Runz,
Kevin Guelton and Herman Akdag

Abstract The design of a fault detection device represents one of the major challenges that manufacturers of robotic systems face today. The detection process requires the use of a number of sensors to monitor the operation of these systems. However, the implementation costs and constraints of these sensors often lead designers to optimize the number used. This could accordingly induce a lack of necessary measures for the optimal detection of failures. One way to bridge this gap consists of realizing model-based estimations of non-measurable state variables describing the dynamics of the real system. This paper presents an approach based on mixed data (measured data and estimated data) for the detection of faults in robotic systems. The proposed fault detection approach is performed using a decision tree classifier. The data used to build this learning stage are obtained from the available measurements of the real system, according to its standard actions. Then, to improve the database classification with unmeasurable data, a linear observer is designed from an analytical model. From the estimations provided by the linear observer, new attributes

R. Taleb · R. Mazouzi · L. Seddiki (✉) · H. Akdag
LIASD EA 4383, University of Paris 8, 2 rue de la Liberté,
93526 Saint-Denis Cedex, France
e-mail: lynda.seddiki@ai.univ-paris8.fr

R. Taleb
e-mail: rtaleb@ai.univ-paris8.fr; Rabih.Taleb@zodiacaerospace.com

R. Mazouzi
e-mail: rabah@ai.univ-paris8.fr

H. Akdag
e-mail: kevin.guelton@univ-reims.fr

C. de Runz · K. Guelton
CReSTIC EA 3804, University of Reims Champagne-Ardenne,
de la Housse, BP 1039, 51687 Reims Cedex 2, France
e-mail: cyril.de-runz@univ-reims.fr

K. Guelton
e-mail: kevin.guelton@univ-reims.fr

R. Taleb
Zodiac Seat Actuation & Control, 139 rue Rateau - Parc des Damiers,
bât. E - CS 70004, 93126 La Courneuve Cedex, France

© Springer International Publishing AG 2018   109
B. Pinaud et al. (eds.), *Advances in Knowledge Discovery and Management*,
Studies in Computational Intelligence 732, https://doi.org/10.1007/978-3-319-65406-5_5

are built, with the aim of enriching the knowledge used by the classifier and thus improving the rate of fault detection. Finally, an experiment on a robotized actuated seat is presented to illustrate the proposed combined linear observer and classifier approach.

1 Introduction

Currently, due to increasing demand for higher system performances and product quality, the complexity of industrial systems is continuously growing. The development of such industrial systems requires the establishment of safe and reliable techniques. To improve the availability, safety and reliability of industrial systems, many methods of supervision, health monitoring, fault detection and isolation (FDI) have been developed. These methods are especially used to detect, isolate and identify any types of fault that may occur in systems such as those used in aircraft, trains and industrial plants to minimize maintenance time and the overall operational costs. Over the past decades, the research community has focused increasingly on this domain and achieved significant progress by developing and implementing many FDI methods, which can be grouped into two main categories: model-based methods (Venkatasubramanian et al. 2003c; Isermann 1984, 2011; Ding 2013; Bouibed et al. 2014) and data-driven methods (Ondel 2006; Venkatasubramanian et al. 2003b; Narvaez 2007).

The first category of FDI methods, model-based methods, is considered when sufficient knowledge of the internal functioning of a system as well as its physical parameters are available (Venkatasubramanian et al. 2003a). These methods can thus accurately represent the functioning of a robotic system to develop soft sensors using state observers. Such state observers can be used as an alternative to physical sensors to reduce the cost of industrial systems and facilitate their mechanical design (Luenberger 1964; Isermann 1984). State observers consist of estimating the internal states (non-measured states) of a system from measurements of its inputs and outputs. The state estimation obtained from observers is most suitable for fault detection and diagnosis thanks to rapid calculations and short time delay in the real-time decision-making process compared with the parameter estimation approach (Zhang and Jiang 2008).

The second category of FDI methods, data-driven methods, is based on historical process data from various sensors available in an industrial system, as well as data from simulation software, during either normal operations or in the case of failure (Vaija et al. 1986). The use of this category of FDI methods is preferred when a non-linear system is considered, whose complexity makes it difficult to obtain an accurate model. This category of FDI methods is widely used in industrial applications. This is because data-driven approaches to detect faults are easy to implement and require very little modeling effort. However, these methods need sufficient knowledge (i.e., a rich database and historical process data) to construct a functioning model, which requires long calculation times, a large storage memory and many sensors (Venkata-subramanian et al. 2003b).

Designing methods that can detect, diagnose and predict faults appears convenient to ensure the good functioning of a robotic system. Nevertheless, one should understand and realize that no single method is adequate to handle all the requirements for FDI systems. Though all methods are restricted and each one has its own advantages and disadvantages, it is obvious that hybrid approaches that combine and merge different methods that work in union to solve different parts of the problem are interesting and attractive. However, very little work addresses such hybrid approaches (Narasimhan et al. 2010; Schuberte et al. 2011).

In Narasimhan et al. (2010), the proposed approach uses analytical models to diagnose electrical, mechanical or thermal faults. These analytical models are combined with a classifier that uses vibration data from the electro-mechanical actuator. The proposed approach extracts data only when necessary to minimize the size of the database. This idea is interesting when the system has a small memory, as in our case study. In addition, the construction of the decision tree is done offline to reduce the computation time, which could influence the quality of the obtained results.

In Schuberte et al. (2011), a unified approach is proposed for FDI. It combines the model-based methods with the data-driven methods. This approach merges model-based methods such as unknown input observer with multivariable statistical process methods to combine the advantages of both techniques and detect and diagnose the faults of linear systems. It should be noted that this approach is not highly efficient when time-varying faults occur, especially for nonlinear systems.

This paper presents an approach based on mixed data (measured data and estimated data) for the detection of faults in robotic systems. This detection is performed using a decision tree classifier type. The data used to build this learning stage are obtained from the available measurements of a real system, according to its standard actions. Then, to improve the database with unmeasurable data, a linear observer is designed from an analytical model. From the estimations provided by the observer, new attributes are built, with the aim of enriching the knowledge used by the classifier and thus improving the rate of fault detection.

The paper is structured as follows: Sect. 2 describes the industrial context. Section 3 develops the proposed approach using real measured and estimated data from a linear state observer. Section 4 presents the application of our approach on a robotic seat and deliver results. Section 5 summarizes the paper.

2 Industrial Context

The goal of this study is to detect and diagnose faults using statistical classification. In this section, we present the industrial context. On one hand, we have a robotic system with various sensors used to measure physical variables and, on the other hand, the linear state observer, which estimates non-measurable variables used as estimated data.

Fig. 1 Robotic system
(actuated seat)

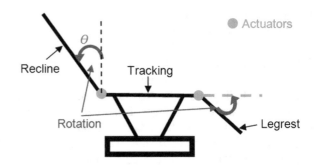

2.1 The Robotic System

This section presents the considered robotic system, which is an actuated seat devoted to be used in a confined space. It can be represented by a robot with tree structure. Indeed, this seat is composed of three actuated bodies (recline, tracking, and leg rest) (cf. Fig. 1). Each body is actuated by an electrical actuator, including a potentiometer that delivers the corresponding angular position. This angular position variates continuously as a result of the control torque applied by the actuator. An electronic control box (ECB) sends commands based on the user's instructions (e.g., forward, backward, stop) to the different actuators to move the different bodies.

This study particularly concerns the recline body, which is modeled as an inverted pendulum actuated by an electrical actuator. This system has different input variables: the electrical supply voltage, the electrical current and the applied torque. It has two kinematic states (variables): the angular position and the angular velocity $(\theta, \dot{\theta})$. Only the angular position is measured because we do not have a velocity sensor. The purpose is then to estimate the angular velocity $(\dot{\theta})$ using the state observer presented below.

2.2 The State Observer

In this section, we present the considered linear state observer (cf. Fig. 2). It consists of estimating the non-measured state (angular velocity) of the considered actuated recline from measurements of the inputs and outputs of the system, namely the applied torque (u) and the angular position (θ) respectively.

To design an observer for the considered actuated seat, it is essential to obtain a model that presents its dynamic behavior. The dynamical model is obtained by the Lagrange formulation, which provides a description of the relationship between the joint actuator forces and the motion of the mechanism and fundamentally operates on the kinetic and potential energy in the system (Khalil and Dombre 1999; Siciliano and Khatib 2008).

Fig. 2 State observer

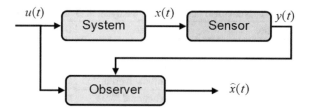

The dynamical model of the recline, considered as an inverted pendulum, is given by:

$$J\ddot{\theta}(t) - mgl \sin \theta(t) = u(t) \tag{1}$$

where u(t) (N m) represents the input torque applied to the recline system by the electrical actuator, θ (rad), $\dot{\theta}$ (rad/s) and $\ddot{\theta}$ (rad/s^2), which are respectively the angular position, velocity and acceleration, and with the parameters given in Table 1.

Note that we have one nonlinearity $\sin \theta(t)$ in the dynamic equation (1). Thus, to design a linear continuous-time observer, a Taylor approximation around the equilibrium point '0' of the dynamic model allows us to derive the linear continuous-time state space model given by:

$$\begin{cases} \dot{x}(t) = Ax(t) + Bu(t) \\ y(t) = Cx(t) \end{cases} \tag{2}$$

where $x(t) = \begin{bmatrix} \theta(t) \\ \dot{\theta}(t) \end{bmatrix}$ is the state vector, $u(t)$ is the input, $y(t)$ is the output and

$A = \begin{bmatrix} 0 & 1 \\ \frac{Mgl}{J} & 0 \end{bmatrix}$, $B = \begin{bmatrix} 0 \\ \frac{1}{J} \end{bmatrix}$ and $C = \begin{bmatrix} 1 & 0 \end{bmatrix}$.

Let us now consider a linear observer given in the following (Luenberger 1964):

$$\begin{cases} \dot{\hat{x}}(t) = A\hat{x}(t) + Bu(t) + L(y(t) - \hat{y}(t)) \\ \hat{y}(t) = C\hat{x}(t) \end{cases} \tag{3}$$

where $\hat{x}(.)$ is the estimated state vector, $\hat{y}(.)$ is the estimated output and L is the observer gain, which is designed to make the continuous-time error dynamics converge to zero asymptotically.

Hence, the gain L is obtained by the pole placement method that consists of placing the poles of the observer in the negative real-part. In our case, we have chosen to place the poles at -5 and -1. With the numerical values of the considered model given in Table 1, one obtains $L = [6\ 215.5]^T$.

The purpose of the use of this observer is to estimate the angular velocity ($\dot{\theta}$) and thus to better detect the faults. Figure 3 shows the angular velocity of the recline model and that estimated by the state observer. The two curves are superposed, which explains the reliability of the estimated velocity by the state observer. Because of this superposition, we decided to use different colors and line-widths of the two curves

Table 1 Numerical parameters of the dynamical model of the recline

Symbol	Parameter	Unit	Value
m	Mass	kg	0.7
l	Length	m	0.33
J	Rotational inertia	$\mathrm{kg\,m^2}$	0.0108
g	Gravitational force	$\mathrm{m\,s^{-2}}$	9.81

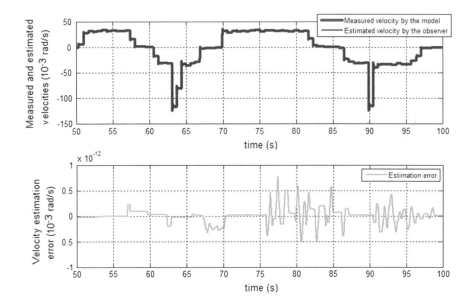

Fig. 3 Angular velocity estimation

to visualize them at the same time. Thus, the velocity of the model is plotted in red while the estimated velocity is plotted in blue. The line-width of the red curve is double of that of the blue curve.

In Fig. 3, the green curve represents the estimation error, which is the difference between the measured angular velocity and the estimated angular velocity from the observer. Along this experiment, the estimation error is small (less than 10^{-12} rad/s), which guarantees the reliability of the velocity estimation. Note that using an inverse model of the inverted pendulum is insufficient to estimate the angular velocity signal since in practice, the initial conditions (initial configuration of the seat) are unknown. A dynamic state observer is convenient since it ensures the asymptotical convergence of the state estimation error.

The linear state observer used in this study being now designed, the next section presents the entire proposed FDI approach, combining the observer-based estimations with classifiers.

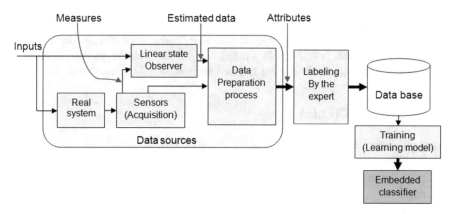

Fig. 4 Block diagram of the proposed FDI approach

3 Proposed Approach

The aim of our approach is to enrich the database of a classifier used in the FDI process by adding estimated data from the state observer designed in the previous section. The estimated angular velocity ($\dot{\theta}$) provides a new attribute named 'estimated velocity'. This enrichment of attributes further aims to increase knowledge on the functioning of the classifier system and thus improve the rate of fault detection.

The block diagram given in Fig. 4 shows the different levels of the proposed approach. It is constituted by five essential blocks:

- The real system (actuated seat) equipped with different sensors providing the measured data,
- The linear state observer which provides the estimations of the non-measured data in this study: the angular velocity),
- The data preparation process,
- The learning model (decision tree),
- The classifier for the FDI.

3.1 Data Preparation Process

In this section, we describe the preparation process of the data which are used in the learning and classification level. The purpose is to build a classification model with the learning step and make predictions for new data.

3.1.1 Measured Data (Acquisition)

The raw data are measured by various sensors installed on the actuated seat; these data represent physical variables and states. They contain details such as time, electrical current and angular position. The sample time of the acquisition is about 300 ms, which is why we have used aggregated data to represent each instance by an entire movement (e.g., rise, fall, stop). It is thus necessary to establish post processing of the measured data to obtain normal/failure functioning indicators.

3.1.2 Attribute Creation

Measurements of the inputs and outputs of the system (actuated seat), respectively the applied torque and the angular position, are inserted in the state observer to estimate the non-measurable state, which, in turn, is used to add a new attribute to the database.

The angular position is one of the most important attributes. The angular position allows the classifier to determine if the movement corresponds to the system inputs (e.g., current, torque, voltage). Another important attribute is the angular velocity. With this attribute, we know whether the system is functioning normally. Some typical faults in actuated systems are mechanical blockades in ball-bearing paths, windings shorts and sensor faults. It is possible to derive analytical relations for the dynamic behavior of the actuated seat. This is why we want to estimate the angular velocity using the state observer to enrich the number of attributes that we have and to increase knowledge on the functioning of the classifier system and thus improve the rate of fault detection.

Another fault is vibration or irregularity during movement. In fact, it is possible that a body of the system (e.g., recline) changes trajectory and breaks down because of vibration. To solve this problem and detect the vibration fault, we added an attribute named 'regularity of movement', which indicates the dispersion of the points representing the angular positions compared to a straight line, which represents an ideal regular movement. This is used as an extra attribute that is calculated after post processing of the measured data (time and angular position), as seen in Fig. 5.

Fig. 5 Movement
(regular/irregular)

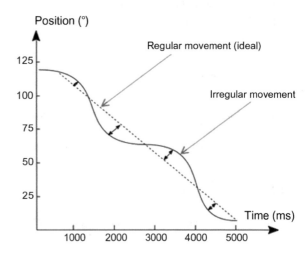

Table 2 Examples of instances

	Movement	Time (ms)	Position (step)	I_i (mA)	I_s (mA)	I_b (mA)	Movement regularity rate	Velocity (10^{-3} rad/s) (estimated)	Class
(a)	01	13730	−399	152	206	2	0.9983	−26.9	N
(b)	02	13507	399	330	250	142	0.9991	28.9	N
(c)	01	5848	−97	140	58	−18	0.9798	−15.1	I
(d)	02	9653	282	310	1100	120	0.9991	28.8	O
(e)	01	8358	−328	140	80	−11	0.9626	−37.6	S
(f)	01	15433	−158	160	126	0	0.6171	9.9	B

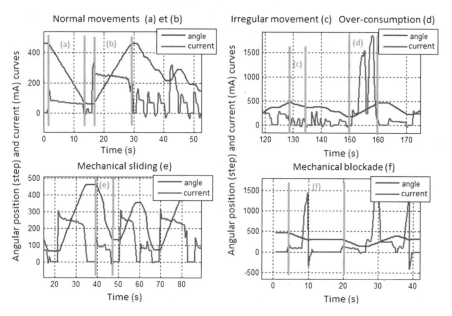

Fig. 6 Position and current for various functioning cases: normal functioning, irregular movement, current overconsumption, mechanical sliding and blockade

Examples of instances with the different attributes of the database are shown in Table 2. These examples are also illustrated in Fig. 6. These attributes and classes denotes:

I_i: Inrush current consumed during the first second of the movement, I_s: Stable current consumed during the permanent operational regime, I_b: Breaking current consumed during the last second of the movement (breaking phase).

N: Normal; I: Irregular; O: Overconsumption; S: Mechanical sliding; B: Mechanical blockade.

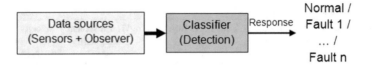

Fig. 7 Classification pipeline

Different functioning cases are shown in Fig. 6 by plotting the curves of the corresponding angular positions (steps) and the electrical current (mA) versus time (s). The two examples 'a' and 'b' show a fall followed by a rise in the normal functioning. Example 'c' shows an irregular fall representing vibration. Current overconsumption is illustrated in example 'd' because the electrical current is about 1700 mA while normal current is about 300 mA. A mechanical sliding fault in example 'e' can be noted due to the bad functioning of the break. This represents a dramatic fall of the measured position in a very short time compared to the other examples. A mechanical blockade is shown in example 'f'. This type of faults ends with a complete stop of the actuator during a period of a few seconds.

3.1.3 Labeling

In this step, the different instances are labeled by the expert responsible for the acquisition phase as he knows the different functioning cases. Thus, labeling decisions are made by the expert in the case of supervised learning.

3.2 The Learning Process and Classification

Before performing the classification step, we must train the classifier using the training data obtained after the data preparation process. In classification studies, it is important to measure classifier performance in terms of the error rate. For our case, we used the classification algorithms implemented in the WEKA[1] software. The classifier predicts the class of each instance (cf. Fig. 7) or, in other words, the functioning case of each movement. If it is correct, this is counted as a success; if it is incorrect, it is an error. The error rate is simply the proportion of errors made over an entire set of instances and measures the overall performance of the classifier (Witten et al. 2011). The classifier triggers an alarm if one instance represents a fault.

---

[1] Weka is a collection of machine learning algorithms for data mining tasks. http://www.cs.waikato.ac.nz/ml/weka.

4 Experimentation

To validate our approach, we realized two series of experiments on the WEKA software:

- The first series of experiments consists in using only the measured data in the learning and classification steps.
- The second series of experiments uses both the measured data and the estimated data by the state observer in the learning and classification steps.

4.1 Training Data

The input to a classification scheme is a set of instances. These instances are the things that are to be classified by classification algorithms. Each instance is an individual example representing a functioning case. Instances are characterized by the values of a set of attributes, such as electrical current, time, humidity, altitude and hydraulic pressure. Our training data is composed of 114 instances as follows:

- 97 instances of normal functioning,
- 4 instances of electrical overconsumption fault,
- 4 instances of mechanical blockade fault,
- 6 instances of mechanical sliding fault,
- 3 instances of vibration or irregular movement fault.

In this application, we used eight attributes, including one estimated attribute, which is angular velocity.

Table 3 shows the experiment results of multi-class classification using our database in the WEKA software. These classes are: normal movement, irregular movement, overconsumption, mechanical sliding and mechanical blockade. In this table, the results are given for the five methods that had the highest accuracy rates. The accuracy is the proportion of instances correctly classified.

Table 3 Classification results

Classifier	Without estimated attribute	With estimated attribute
	Accuracy rate (%)	Accuracy rate (%)
BayesNet	91.2	92.1
Decision table	90.3	92.1
J48	91.2	92.9
JRip	90.3	93.8
Random tree	88.5	93.8

4.2   Classification Algorithms

There are multiple categories of classifiers (Witten et al. 2011; Liu 2011): decision trees, rules algorithms, Bayes algorithms, neural networks, nearest neighbors algorithms and bagging, among others. In this study, one considers the following classification algorithms:

- J48 which is the implementation of the decision table C4.5 (Quinlan 1993),
- Random Tree (Witten and Frank 2005),
- BayesNet (Bouckaert 2004),
- Decision table (Kohavi 1995),
- JRip (Fürnkranz and Widmer 1994).

C4.5 and Random Tree are algorithms used to generate decision trees that can be used for classification. Decision tree consists of checking for base cases, calculating the gain ratio for each attribute and choosing the one with the highest normalized information gain, then creating a decision node that splits the database and does the same for the children of nodes. In the case of Random Tree, the tree is defined by linear regression and provides a piecewise linear regression model while C4.5^2 gives a piecewise constant model.

BayesNet classifier is a Bayesian network that produces probability estimates rather than hard classifications. It consists of estimating the probabilities that a given instance belongs to each class value.

Finally, JRip and Decision table are two classification rules algorithms using a 'covering' approach that consists in identifying, at each stage, a rule that covers some of the instances. Decision table uses a table of decisions to represent the rules.

4.3   Classification Parameters

For the purpose of measuring the performance of classification algorithms, we divide the used database into a training set and a test set. Different parameters can be used for this purpose. It is important that the test dataset is not used in any way to create the classification model. Witten et al. (2011) explained that in many situations, people talk about three datasets: the training dataset, the validation dataset and the test dataset. The validation data are used to optimize parameters of the classifiers or to select a particular one.

Because the amount of data in the database is limited, we decided to divide the database into two data sets (training and test) and choose the cross-validation parameter for training and testing phases. Generally, this consists of dividing the database into 10 parts. Each part is held out in turn and the remaining nine-tenths are used for the training phase to create the functional model. This last is then used for testing the retained part, and the error rate is calculated. This procedure is repeated 10 times on different training sets. Finally, the final value of the error rate is the

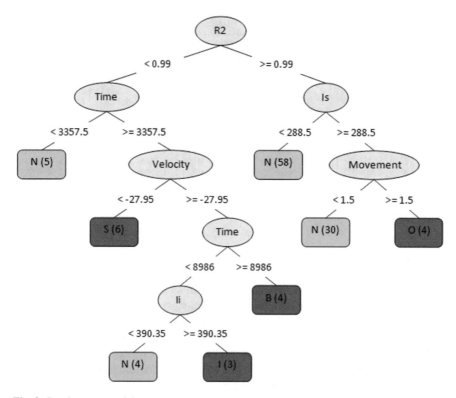

Fig. 8 Random tree model

average of the 10 error rates previously calculated. In our case, we used stratified cross-validation to ensure that each part is a representative of the whole database. More details are explained in Witten et al. (2011) and Liu (2011). Figure 8 shows the tree model of the 'Random Tree' algorithm used and demonstrates the complexity of the model.

4.4 Results

After simulating our database with the algorithms mentioned above, we obtained the results in Table 3. These results show that regardless of the method used, the accuracy of classification is improved after adding the estimation attribute by the linear state observer, which is the estimated angular velocity.

The minimum gain of about 0.8% is reached by BayesNet, the maximum gain of about 5.2% is reached by Random Tree, and the mean gain is about 2.6%. The FDI is thus more efficient, which explains why the estimated data enrichment by

Fig. 9 Confusion matrices

Without the estimated attribute

BayesNet

N	V	O	S	B	
93	1	1	2	0	N
1	0	0	1	1	V
0	0	3	0	1	O
0	1	0	5	0	S
0	0	1	0	3	B

Decision table

N	V	O	S	B	
96	0	0	1	0	N
3	0	0	0	0	V
1	0	3	0	0	O
4	0	0	2	0	S
2	0	0	0	2	B

J48

N	V	O	S	B	
96	0	1	0	0	N
2	0	0	1	0	V
0	0	4	0	0	O
1	3	0	2	0	S
0	0	1	1	2	B

JRip

N	V	O	S	B	
94	2	0	1	0	N
2	0	0	1	0	V
1	0	3	0	0	O
2	1	0	3	0	S
1	0	0	0	3	B

Random tree

N	V	O	S	B	
93	4	0	0	0	N
2	0	0	1	0	V
2	0	2	0	0	O
1	2	0	3	0	S
0	0	0	1	3	B

With the estimated attribute

BayesNet

N	V	O	S	B	
94	1	1	1	0	N
1	0	0	1	1	V
0	0	3	0	1	O
0	1	0	5	0	S
0	0	1	0	3	B

Decision table

N	V	O	S	B	
96	0	0	1	0	N
3	0	0	0	0	V
1	0	3	0	0	O
2	0	0	4	0	S
2	0	0	0	2	B

J48

N	V	O	S	B	
95	0	1	1	0	N
3	0	0	0	0	V
0	0	4	0	0	O
1	0	0	5	0	S
0	1	1	0	2	B

JRip

N	V	O	S	B	
96	0	0	1	0	N
3	0	0	0	0	V
1	0	3	0	0	O
1	0	0	5	0	S
1	0	0	0	3	B

Random tree

N	V	O	S	B	
96	1	0	0	0	N
2	0	0	1	0	V
1	0	3	0	0	O
0	2	0	4	0	S
0	0	0	0	4	B

the state observer provides real improvements for the functioning operation of the robotic system (actuated seat).

We can use one of the two best classifiers with an accuracy about 93.8%. In our case, we chose Random Tree because it has the highest gain (5.3%) compared to the results before and after adding the estimated attribute. According to PIPAME[2] and ICF International,[3] the Maintenance, Repair and Overhaul (MRO) cost of equipment in the civil aircraft field reached $13.7 billion in 2014. Our approach may at least reduce this cost by limiting the number of unnecessary maintenance operations of the real system. Random Tree performs a reduction of 5.3%.

However, this classifier has an error rate of about 6.2% which is a significant proportion in some industrial areas. A detailed study was done for each value of the class using confusion matrices. A confusion matrix is used in the case of multi-class classification and prediction. It is often displayed as a two-dimensional matrix with a row and column for each class. All instances of the test dataset are used, which means that we used the entire database composed of 114 instances. The row represents the actual class and the column represents the predicted class.

Detailed confusion matrices are presented in Fig. 9. This comparison study shows that normal class 'N' is very well classified thanks to the sufficient number of instances of this class while vibration class 'V' is poorly classified either because of the limited number of instances or the fact that vibration functioning is highly similar to normal functioning. This explains why at least one vibration instance is classified as normal functioning in each test.

The classification results of mechanical sliding 'S' are improved by adding the estimated attribute. In four out of five cases, the recall of 'S' increased by 40% on average. We can conclude that this approach is adequate for fault detection.

Adding the estimated attribute improves the fault detection results, which indicates why the use of the proposed approach is relevant. The design of new and more efficient

---

[2]http://www.economie.gouv.fr/.

[3]http://www.icfi.com/.

methods of classification can thus improve the classification results. For this reason, the use of better estimation methods such as nonlinear observers and unknown input observers should be contemplated in further prospects.

5 Conclusions and Future Work

This article, which presents a diagnosis approach for fault detection, supports the fusion of model-based methods and data-driven methods. This approach consists of the use of mixed data (measured and estimated) for fault detection in robotic systems. This detection is performed using a decision tree classifier. Data is measured from the real system using physical sensors. This data is then enriched with estimated data from a linear state observer to increase knowledge of the functioning of the classifier system and thus improve the rate of fault detection.

We also present the application of this approach for an actuated seat, showing its advantages in the detection of failure. The addition of the estimated attribute improves the classification results and thus the fault detection. This means that we can reduce the number of physical sensors in a robotic system by employing software sensors and estimating the corresponding variables. By reducing the number of sensors, we can reduce maintenance time and overall operational costs, especially the cost of sensors and their implementation.

In the future, we would like to test this approach on other robotic systems and other types of databases to generalize it. In addition, we plan to use the nonlinear model of the inverted pendulum and nonlinear state observers to improve the precision reached due to adding the estimation and predict some failure. This prediction can be done by modeling the three bodies of the actuated system and setting up multiple diagnosis systems.

Acknowledgements The authors would like to acknowledge the contributions of their colleagues and collaborators at Zodiac Seat Actuation & Control (ZSAC). Financial support for this work was provided by ZSAC.

References

Bouckaert, R. R. (2004). Bayesian network classifiers in weka. Technical report, University of Waikato, New-Zealand.

Bouibed, K., Seddiki, L., Guelton, K., & Akdag, H. (2014). Actuator and sensor fault detection and isolation of an actuated seat via nonlinear multi-observers. *Systems Science & Control Engineering: An Open Access Journal, 2*(1), 150–160.

Ding, S. X. (2013). *Model-Based Fault Diagnosis Techniques: Design Schemes*. Algorithms and Tools: Springer Science & Business Media.

Fürnkranz, J., & Widmer, G. (1994). Incremental reduced error pruning. In W. W. Cohen & H. Hirsh (Eds.), *Machine learning, proceedings of the eleventh international conference, Rutgers University, New Brunswick, NJ, USA, July 10–13, 1994* (pp. 70–77). Morgan Kaufmann.

Isermann, R. (1984). Process fault detection based on modeling and estimation methods–A survey. *Automatica, 20*(4), 387–404.

Isermann, R. (2011). *Fault-diagnosis applications: model-based condition monitoring: actuators, drives, machinery, plants, sensors, and fault-tolerant systems*. Springer Science & Business Media.

Khalil, W., & Dombre, E. (1999). *Modélisation, identification et commande des robots*. Hermès science publ.

Kohavi, R. (1995). The power of decision tables. In *Proceedings of the European conference on machine learning* (pp. 174–189). Springer.

Liu, B. (2011). *Web Data Mining: Exploring Hyperlinks, Contents, and Usage Data. Data-Centric Systems and Applications* (2nd ed.). Springer.

Luenberger, D. G. (1964). Observing the state of a linear system. *IEEE Transactions on Military Electronics, 8*(2), 74–80.

Narasimhan, S., Roychoudhury, I., Balaban, E., & Saxena, A. (2010). Combining model-based and feature-driven diagnosis approaches—A case study on electromechanical actuators. In *21st international workshop on principles of diagnosis* (pp. 1–8).

Narvaez, C. V. I. (2007). *Diagnostic par techniques d'apprentissage floues: concept d'une méthode de validation et d'optimisation des partitions*. PhD thesis, INSA de Toulouse.

Ondel, O. (2006). *Diagnostic par reconnaissance des formes: Application à un ensemble convertisseur-machine asynchrone*. Ph.D. thesis, Ecole Centrale de Lyon.

Quinlan, J. R. (1993). *C4.5: Programs for machine learning*. San Francisco, CA: Morgan Kaufmann Publishers Inc.

Schubert, U., Kruger, U., Arellano-Garcia, H., de Sá Feital, T., & Wozny, G. (2011). Unified model-based fault diagnosis for three industrial application studies. *Control Engineering Practice, 19*(5), 479–490.

Siciliano, B., & Khatib, O. (2008). *Springer handbook of robotics*. Springer Science & Business Media.

Vaija, P., Järveläinen, M., & Dohnal, M. (1986). Failure diagnosis of complex systems by a network of expert bases. *Reliability Engineering, 16*(3), 237–251.

Venkatasubramanian, V., Rengaswamy, R., & Kavuri, S. N. (2003a). A review of process fault detection and diagnosis: Part II: Qualitative models and search strategies. *Computers & Chemical Engineering, 27*(3), 313–326.

Venkatasubramanian, V., Rengaswamy, R., Kavuri, S. N., & Yin, K. (2003b). A review of process fault detection and diagnosis: Part III: Process history based methods. *Computers & Chemical Engineering, 27*(3), 327–346.

Venkatasubramanian, V., Rengaswamy, R., Yin, K., & Kavuri, S. N. (2003c). A review of process fault detection and diagnosis: Part I: Quantitative model-based methods. *Computers & Chemical Engineering, 27*(3), 293–311.

Witten, I. H., & Frank, E. (2005). *Data mining: Practical machine learning tools and techniques (Morgan Kaufmann series in data management systems)* (2nd ed.). San Francisco, CA: Morgan Kaufmann Publishers Inc.

Witten, I. H., Frank, E., & Hall, M. A. (2011). *Data mining: Practical machine learning tools and techniques* (3rd ed.). San Francisco, CA: Morgan Kaufmann Publishers Inc.

Zhang, Y., & Jiang, J. (2008). Bibliographical review on reconfigurable fault-tolerant control systems. *Annual Reviews in Control, 32*(2), 229–252.

Author Biographies

Rabih Taleb is a Ph.D. Candidate in computer science at the University of Paris VIII within the framework of cooperation between LIASD of Paris VIII and Zodiac Aerospace. He received his Master's degree in Mechatronics systems from the engineering school of the university of Orléans,

France. His main research interests are fault detection and diagnosis, health monitoring and fault prognosis in industrial and robotic systems.

Rabah Mazouzi received the Ph.D. degree in computer science in 2016 from the university of Paris VIII. He is currently working as a temporarily associated to teaching and research in the department of information and communications technology of Paris VIII university. His research interests include data mining, cloud computing and big data analysis in the field of machine learning in particular data classification.

Lynda Seddiki received the Ph.D. degree in robotic and automatic control at the University of Reims Champagne Ardenne (URCA), Reims, France, in 2008. She is currently an Associate Professor at the University of Paris VIII. Since 2009, she carries out research in LIASD Laboratory at Saint-Denis, France. His main research interests are Robotics systems,Takagi-Sugeno models, robust control and diagnosis.

Cyril de Runz is a lecturer at the University of Reims Champagne-Ardenne. He obtained his Ph.D. in computer science in 2008 from the same university. His fields of interests are Artificial Intelligence, Fuzzy Set Theory, Geomatics, Data Mining and Information Systems. He organized several workshops, special sessions and conferences and published more than 50 papers.

Kevin Guelton is an Associate Professor in automatic control at the University of Reims Champagne Ardenne and a researcher at the CReSTIC. His major field of study includes Takagi-Sugeno and hybrid systems control and observation. Kevin Guelton is a member of the IFAC TC 3.2 and TC 8.2, served as general chair of LFA 2013 and IFAC ICONS 2016 and in more than 40 IPC of international events. He is an editor of engineering Application of Artificial Intelligence.

Herman Akdag is presently full Professor of computer science at Paris VIII University. Initially focused on Information Theory and Knowledge Representation, his scientific orientations evolved to Machine Learning and Cognitive Modeling via Fuzzy Logic. He is actually member of LIASD Lab, the head of Master CPI, and member of the University Paris VIII Senat until May 2020. He published more than 170 papers in conferences proceedings, journals and books.

A Collaborative Framework for Joint Segmentation and Classification of Remote Sensing Images

Andrés Troya-Galvis, Pierre Gançarski and Laure Berti-Équille

Abstract In this article, we present a collaborative framework for joint segmentation and classification. The framework is guided by and aware of the quality of each segment at every stage; it allows the consideration of both homogeneity based criteria as well as implicit semantic criteria to extract the objects belonging to a given thematic class. We apply the proposed framework to vegetation extraction in a very high spatial resolution image of Strasbourg. We compare our results to a pixel-based method, an object-based method and a hybrid segmentation-classification method. The experiments show that the proposed method reaches good classification results while remarkably improving the segmentation results.

1 Introduction

Automatic interpretation of remote sensing images is a difficult but crucial task in a wide range of applications. Since the apparition of Very High Spatial Resolution (VHSR) imagery, the Object Based Image Analysis (OBIA) paradigm has been preferred over pixel oriented approaches (Blaschke 2010). Indeed, at VHSR a single pixel is not informative enough for the classification task since the types and complexity of identifiable objects increase considerably. Thus, image segmentation is performed first in order to obtain higher level objects called segments which allow a better description of the image.

Figure 1 illustrates the complete workflow of OBIA. First (V)HSR images are acquired via satellite sensors. Atmospheric and geometric corrections are applied as pre-processing to produce ready-for-analysis images. Segmentation is then applied

A. Troya-Galvis (✉) · P. Gançarski
iCube UMR CNRS 7357, University of Strasbourg, Strasbourg, France
e-mail: troyagalvis@unistra.fr

P. Gançarski
e-mail: gancarski@unistra.fr

L. Berti-Équille
Qatar Computing Research Institute, Hamad Bin Khalifa University, Doha, Qatar
e-mail: lberti@qf.org.qa

© Springer International Publishing AG 2018   127
B. Pinaud et al. (eds.), *Advances in Knowledge Discovery and Management*,
Studies in Computational Intelligence 732, https://doi.org/10.1007/978-3-319-65406-5_6

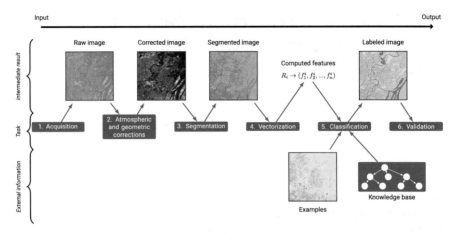

Fig. 1 Typical workflow of OBIA methods

and the image is now described by segments, which are sets of similar pixels satisfying a given homogeneity criterion. A vectorization stage consisting in representing each segment as a vector of descriptive features such as shape, textural, or sophisticated radiometric indexes is then applied. Next, classification techniques which may use external information such as examples or ontologies, are applied to obtain a full land-cover map of the image. As a final step, the resulting image is validated either by an expert or by objective metrics in order to assess the quality of the results.

Segmentation and classification are—in our opinion—the most critical steps. Many classification techniques such as (di Sciascio et al. 2013), have proven to be successful when the segmentation approaches a one-to-one mapping between the segments and geographic objects. Unfortunately, there is no single segmentation algorithm leading to such perfect mapping in all cases. Our intuition suggests that (1) the improvement of image segmentation should lead to the improvement of the classification; and (2) classification results may be useful to improve the quality of the segmentation. Moreover, while full interpretation is often desirable, the extraction of a single thematic class is generally sufficient for multiple tasks such as forest mapping (Räsänen et al. 2013), deforestation tracking (Duveiller et al. 2008), landslide risk management (Promper et al. 2014), urban planning (Pham et al. 2011), etc. Thus, we propose a collaborative framework for joint segmentation and classification (CoSC) which is quality-centric and attempts to simultaneously improve both segmentation and classification for a given thematic class by the interaction of these two paradigms which are closely related in the OBIA context.

The rest of this article is structured as follows: Sect. 2 introduces the context and related work. Section 3 presents our collaborative framework for joint segmentation and classification. Section 4 presents an experimental study demonstrating the applicability of the proposed framework. Finally, in Sect. 5 we conclude and give some research perspectives.

2   Related Work

Many work can be found in the literature about combining image segmentation and classification methods in a remote sensing context. For instance, (Lizarazo and Elsner 2011) proposed a fuzzy segmentation approach in which the image pixels are classified by fuzzy classifiers trained for every desired class, thus obtaining what they call fuzzy segments; the fuzzy segments are defuzzyfied and merged together following a set of logical rules in order to obtain a fully labelled image. Derivaux et al. (2010) proposed a supervised segmentation algorithm which adapts the parameters of the watershed segmentation algorithm by exploiting examples given by an expert and applying evolutionary algorithms. Kurtz et al. (2012) developed a hierarchical multi-resolution approach for the extraction of complex urban objects. They take advantage of multi-source images which allows to have different views on the same data; using low resolution images to extract coarse objects such as a whole city and higher resolution images to extract finer objects like a district or individual buildings. Mahmoudi et al. (2013) proposed a multi-agent system for remote sensing image classification where each agent is specialized in the extraction of a given class. A particular agent is charged to manage conflict resolution. Tarabalka et al. (2009) presented Spectral-Spatial Classification (SSC), a hybrid approach mixing pixel- and object-based approaches; it uses EM clustering as segmentation algorithm, then a pixel-wise SVM classification is performed and a voting scheme is employed to decide the label for each segment. A final filtering step is applied to remove small noisy segments. Recently, (Hofmanna et al. 2014) formalised an Agent Based Image Analysis (ABIA) framework, in which the idea is to let every segment be an independent agent which modifies itself trying to improve and modify classification rules given by a domain ontology.

Our work is mainly inspired by the wrapper-based segmentation framework (Farmer 2009), which aims at segmenting complex objects by the integration of a semantic context to the segmentation process. This integration is done by wrapping a classifier inside a segmentation algorithm and use it as a quality metric to optimize. First, an initial segmentation is computed, this segmentation has to be over-segmented in order to the wrapper based framework to be effective. Indeed, the further steps consist in the filtering of irrelevant segments, the remaining segments are likely to be part of the object of interest. Then an optimization algorithm which alternates a segmentation step (adding or removing a segment) with a classification step. When the classification probability reaches an acceptable threshold, the best segmentation found and the classified objects are returned. The use of the classification to guide the segmentation process eliminates the homogeneity constraint on the segmentation allowing the extraction of complex objects.

Nevertheless, the wrapper framework supposes that the number of objects to extract is known to be relatively low and that their position can be easily determined to filter out unnecessary regions. These assumptions are generally not satisfied within a remote sensing context. Indeed, in a real-life scenario, objects of interest may vary from a few tens to a many hundred, or may not be present at all in the image. Moreover, objects of a same class may vary in shape and sizes and can be all over the image so it is not trivial to determine their possible locations.

We propose a generalization of the wrapper framework which allows the modification of segments in a more flexible manner making possible to extract objects of a given class regardless of their shape or position.

3 Proposition

The segmentation and classification paradigms pursuit different goals, yet they are closely related in a remote sensing context. Indeed, segmentation tries to partition the image spatially, mainly based on color properties (radiometric responses); while classification aims at partitioning data based on some knowledge which can be explicitly modelled or implicitly injected as training examples. Our intuition is that proper interactions between both approaches should lead to the improvement of both segmentation and classification results simultaneously, as the ideal segmentation is the one allowing to best classify the image and the perfect classification result in a partitioning with a one-to-one mapping between image regions and geographic objects.

Studies have shown the benefits of treating a single thematic class at a time (Musci et al. 2013). In fact, it is easier to set the parameters of a segmentation algorithm so that the resulting segments fit one class without taking care of the rest. As a direct consequence, the classification of objects from this class is improved. Thus, our collaborative framework CoSC is devoted to the extraction of the objects belonging to one thematic class from the image. Let C be such a class of interest, we assume the existence of the following elements:

- An initial segmentation $S = \{R_i \mid 0 < i < W\}$ where W is the number of segments; and where each segment $R_i = \{(x_k^i, y_k^i) \mid 0 < k < M\}$ with M is the number of pixels in R_i.
- A 1-versus-all classifier $\mathbb{C}_C$ allowing to discriminate objects of class C from the rest. Note that this classifier has to be properly trained offline avoiding extreme over- and under-fitting since no further learning is performed during the CoSC process.

3.1 Definitions

Let us pose some definitions which are necessary in order to formally describe the proposed framework.

Definition 1 A specialized segmenter, noted S_C, is a segmentation agent capable of locally evaluating and modifying a given segment R.

Definition 2 A class extractor, noted C_C, is a classification agent capable of determining by using $\mathbb{C}_C$, the probability $P_C(R)$ that a given segment $R \in S$ belongs to the class C.

Definition 3 A CoSC agent, noted SC_C, is an agent capable of managing the collaboration between a specialized segmenter and a class extractor.

Definition 4 Let $T_\in$ (resp. $T_{\notin}$) be a threshold over the probability of being (resp. not being) a C class object. Thus, $T_\in$ and $T_{\notin}$ define a reject zone (Chow 1970) for the classifier $\mathbb{C}_C$.

Definition 5 An ambiguous segment is a segment R such that $T_{\notin} \leq P_C(R) \leq T_\in$.

3.2 Global CoSC Process

The collaborative process is described in Algorithm 4. After the selection of an initial segmentation, the process involve interactions between classification and segmentation approaches. In fact, the following steps are iterated until convergence:

1. A segment selection step relying on a classification-based criterion.
2. A segment modification step relying on local segmentation evaluation and segmentation operators.
3. An evaluation step based on classification probability.

Thus, the first step aims at selecting a poorly classified segment; the second step attempts to correct segmentation errors around the selected segment which in turn should improve its classification; finally, the evaluation step allows the process to quantify the improvement (or deterioration) of the current solution and checking for convergence. In the following subsection we describe in detail one possible implementation of these steps.

3.2.1 Initialization

S_C begins with an initial segmentation S which can be arbitrarily chosen. Nevertheless, over-segmentation is preferable to under-segmentation, since the latter is harder to correct. Our experiments show that the closer the size of the segments are close to the size of C class objects, the quicker collaboration converges.

3.2.2 Candidate Segment Selection

C_C selects a candidate segment R_a to be modified. It is essential to chose a good candidate segment, in other words, a segment which may lead to the improvement of the classification after its modification. Different strategies can be used, such as

---

Algorithm 4: CoSC process

    Input   : Specialized Segmenter: S_C,
                  Class extractor: C_C
    Output: Segmentation: S_{best}
1 **begin**
2     $S \leftarrow S_C.\text{initialSegmentation}()$
3     $S_{best} \leftarrow S$
4     $S_{current} \leftarrow S$
5   **while not** $C_C.\text{convergence}()$ **do**
6         $R_a \leftarrow C_C.\text{candidate}(S_{current})$
7         $R_m \leftarrow S_C.\text{SegmentModifications}(R_a)$
8   Update $S_{current}$ with R_m
9         $C_C.\text{evaluate}(S_{current})$
10   **if** $S_{current}$ is better than S_{best} **then**
11 | Update S_{best} with $S_{current}$
12
13   **return** S_{best}
14 **end**

---

selecting a random segment, or selecting the segment which is closest to $T_\in$ or $T_\notin$. In our experiments (Sect. 4) we select the most ambiguous segment, as follows:

$$R_a = \arg\min_{R_i} |P_C(R_i) - T_{avg}| \tag{1}$$

where $T_{avg} = \frac{T_\in + T_\notin}{2}$.

3.2.3 Segment Modification

S_C evaluates the quality of the candidate segment and then tries to improve it's segmentation consequently. Following a given quality criterion, S_C associates a segment with one of the following states: over-segmentation, under-segmentation or undetected segmentation error.

A segment modification operator is a function $O : \mathfrak{D}_i \to \mathfrak{D}_i$ where $\mathfrak{D}_i = R_i \cup N_{R_i}$ and N_{R_i} denotes the set of points belonging to the segments which are adjacent to R_i (i.e., neighbour segments). Thus, a list containing at least one function O is associated to each segmentation state. For over-segmentation, the list O_L should contain operators which may make the segment bigger. For under-segmentation, the list U_L should contain operators which may make the segment smaller. For undetected segmentation error, the list M_L should contain a variety of mixed operators which may help to improve the classification of the segment. The modifications are applied as depicted in Algorithms 5 and 6. A random operator is applied from one of the three lists to the candidate segment in function of its quality state.

In our implementation (Sect. 4) three different O functions are used.

The first one, is intended to correct heavy over-segmentation, it is a merging operator, noted FN, which merges segment R_i with its closest neighbour R_j:

$$Z = \arg \min_{R_j \in N_{R_i}} \delta_{L*a*b}(R_i, R_j) \tag{2a}$$

$$FN(R_i) = (R_i \cup Z) \cup (N_{R_i} \setminus Z) \tag{2b}$$

where $\delta_{L*a*b}(R_i, R_j)$ is the Euclidean distance between segments R_i and R_j in the $L*a*b$ colorspace (Chen and Wang 2004).

The other two, are intended to correct slight over- and under-segmentation, they are based on morphological operators which may be easily applied to grow or shrink a segment.

The growing operator, noted Gr, is defined by:

$$Gr(R_i) = R_i \oplus \boxdot_1 \cup N_{R_i}^- \tag{3}$$

where $R_i \oplus \boxdot_1$ denotes the morphological dilation (Haralick et al. 1987) of R_i by a 3×3 square; and $N_{R_i}^- = N_{R_i} \setminus ((R_i \oplus \boxdot_1) \setminus R_i)$.

The shrinking operator, noted Sh, is defined by:

$$Sh(R_i) = R_i \ominus \boxdot_1 \cup N_{R_i}^+ \tag{4}$$

where $R_i \ominus \boxdot_1$ denotes the morphological erosion (Haralick et al. 1987) of R_i by a 3×3 square; and $N_{R_i}^+ = N_{R_i} \cup (R_i \setminus (R_i \oplus \boxdot_1))$. Exceeding pixels resulting from the erosion are allocated to the closest neighbour $R_j \in N_{R_i}$.

Note that if the application of an operator O implies modifying the topology of resulting segments, then the segment is returned unmodified.

We set $O_L = \{FN(R_i)\}$; $U_L = \{Sh(R_i)\}$; and $M_L = \{Gr(R_i), Sh(R_i), FN(R_i)\}$. In this case the candidate segment modification strategy is equivalent to the following rules:

- If R_a is over-segmented, it is merged with its most similar neighbour.
- If R_a is under-segmented, it is shrunk by an morphological erosion.
- Otherwise, three operations (merging, growing and shrinking) are tested randomly and the first successful (i.e., actually modifying R_a) transformation is kept.

3.2.4 Evaluation

C_C evaluates the quality of the current solution $S_{current}$ by using an objective function based on classification criteria. Thus, under the assumption that the classifier is well trained and capable of discriminating C class objects from other kind of objects, we can evaluate the solution by looking at how well the classifier separates objects of

---

Algorithm 5: ApplyModifications

Input   : Segment: R_a,
             List of O: L
Output: Segment: R_m
1 **begin**
2 | $Shuffle(L)$
3 | **while** $L.iterator.has\_next()$ **do**
4 | | $R_m \leftarrow modify(R_a, L.iterator.next())$
5 | | **if** $R_m \neq R_a$ **then**
6 | | | **return** R_m
7 | |
8 | **return** R_a
9 **end**

Algorithm 6: SegmentModification

Input   : Segment: R_a,
Output: modified Segment: R_m,
1 **begin**
2 | **switch** $LocalSegmentEvaluation(R_a)$ **do**
3 | | **case** $oversegmentation$
4 | | | $R_m \leftarrow ApplyModifications(R_a, O_L)$
5 | | **case** $undersegmentation$
6 | | | $R_m \leftarrow ApplyModifications(R_a, U_L)$
7 | | **otherwise**
8 | | | $R_m \leftarrow ApplyModifications(R_a, M_L)$
9 | |
10 | **return** R_m
11 **end**

---

interest from the rest. In other words, the goal is to reduce the number of ambiguous segments (c.f., Definition 5). The evaluation function Q_{cs} can be defined as follows:

$$Q_{cs} = \frac{1}{W} \left(\sum_{i \mid P_C(R_i) > T_\in} P_C(R_i) + \sum_{i \mid P_C(R_i) < T_\notin} (1 - P_C(R_i)) \right) \tag{5}$$

where W is the total number of segments in the current segmentation. The idea behind this formula is to reward those segments for which the classifier has made a decision proportionally to the confidence of the classifier. The more the classifier is sure about the class of a given segment R_i, the more R_i is contributing to the score. Plus, ambiguous segments are penalized as they do not contribute to the score, but they are taken into account by the $\frac{1}{W}$ factor. Thus, in the ideal case where the classifier has perfectly classified all of the segments with 100% confidence then $Q_{cs} = 1.0$; in the worst case in which all of the segment are ambiguous then $Q_{cs} = 0.0$. At each iteration S_{best} is updated if $S_{current}$ has a higher Q_{cs} score.

3.2.5 Convergence

A simple strategy is to let the collaboration continue as long as it improves the Q_{cs} score, but it would probably lead to a premature convergence problem, thus to a suboptimal result. To address this issue a variant of simulated annealing is used to drive the optimization process.

3.3 Output

The output of each iteration of a modified segmentation S_{best} as well as a probability image associated to S_{best} in which each pixel $(x_k^i, y_k^i) \in R_i$ correspond to the probability $P(R_i)$.

It is then possible to evaluate these results in terms of classification, or in terms of segmentation as we show in Sect. 4.2.

4 Case Study

The performance of the classifier $\mathbb{C}_C$ has a considerable impact on the results of CoSC. To limit the bias induced by $\mathbb{C}_C$, we chose to apply CoSC for the extraction of vegetation zones. Indeed, in this case we could easily define and train a well-performing classifier, allowing us to better validate our proposition.

The studied image is a 9211×11275 (10^8 pixels) VHSR image of Strasbourg. It is a Pleiades pansharpened image at 50 cm resolution, with 4 spectral bands: red (R), green (G), blue (B), and near infra red (NIR). Figure 2a shows the studied image in false colors (i.e., the red was replaced by the NIR, as vegetation is known to have high radiometric responses in this region of the spectrum). For ease of computation, the image was cropped into 1620 tiles of 256×256 pixels. Each tile was processed with the same parameters. In this paper we analyse in detail the results of three tiles: the first one corresponds to a zone of industrial facilities; the second one corresponds to an urban area; and the third one corresponds to a public park. They are labelled A, B and C respectively on Fig. 2a and shown in detail on Fig. 3b, c. The resulting probability images are compared to available reference-data shown in Fig. 2b.

4.1 Instantiation

The framework was instantiated as follows.

(a)   **(b)**

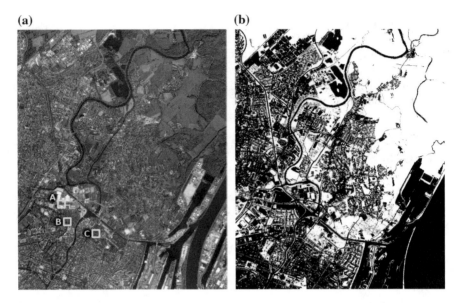

Fig. 2 *Left* studied image. *Right* available reference data

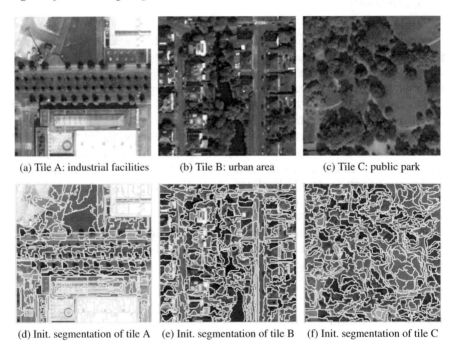

(a) Tile A: industrial facilities   (b) Tile B: urban area   (c) Tile C: public park

(d) Init. segmentation of tile A   (e) Init. segmentation of tile B   (f) Init. segmentation of tile C

Fig. 3 *Top* tiles A, B and C at a detailed scale. *Bottom* the initial mean-shift segmentations computed for those tiles

4.1.1 Specialized Segmenter (Definition 1)

The UOA_{L_2} metric (Troya-Galvis et al. 2015) was used as segmentation quality metric; the homogeneity index used was the entropy (Zhang et al. 2003) and the threshold value was learn from some explicitly given examples of vegetation. The local evaluation function of UOA is used as well to determine whether a segment is over-, under- or well-segmented.

The mean-shift algorithm was applied to generate the initial segmentations (Fig. 3d–f). Its parameters were selected by evaluating 100 different parameter combinations with the UOA_{L_2} metric. Remark that it is possible to apply any other method to select the initial segmentation. Different initial segmentations would surely lead to different results; however a complete sensitivity analysis on this aspect is out of the scope of this article.

4.1.2 Class Extractor (Definition 2)

A linear regression model was learnt using the Weka API (Hall et al. 2009). The training set is a subset of the fuzzy reference data containing both vegetation and non-vegetation segments chosen randomly. The feature set employed was composed of 32 features, including radiometrical attributes such as mean and standard deviation values for each band (Rougier and Puissant 2014), as well as geometrical attributes such as the area, shape, orientation and solidity indexes among others. When the predicted values are out of the accepted bounds, they are simply truncated to 1.0 or 0.0 as needed.

Remark that the definition of the classification thresholds can be misleading. Indeed, the classifier is a parameter of CoSC and it has to be trained offline. Thus, the classification thresholds (Definition 4) within CoSC are merely a tool to "push" the algorithm to look for very confident classifications. Then, these thresholds should have little or no impact on the final classification results. In order to verify this fact, we conducted the experiment for $T_\in = 0.9$, $T_{\notin} = 0.1$; $T_\in = 0.8$, $T_{\notin} = 0.2$; and $T_\in = 0.7$, $T_{\notin} = 0.3$, using the same initial segmentations. The results are summarized in Table 1, we can see that the choice of the classification thresholds has almost no impact on the classification results, however they have some influence on the computation times. In Sect. 4.2 we study in detail the results for $T_\in = 0.9$ and $T_{\notin} = 0.1$.

Table 1 Quality measures for different $T_\in$ and $T_{\notin}$ values

$T_\in$	$T_{\notin}$	Acc	Pr	Re	F1	κ	Am%	Q_{cs}
0.7	0.3	0.932	0.952	0.945	0.949	0.846	0.469	0.526
0.8	0.2	0.930	0.952	0.943	0.948	0.844	0.466	0.529
0.9	0.1	0.930	0.952	0.943	0.947	0.844	0.464	0.531

Table 2 Impact of CoSC parameters on CPU time, segmentation and classification quality

Parameter	CPU time	Segmentation	Classification
$T_\in$	Medium	Very low	Very low
$T_\notin$	Medium	Very low	Very low
S	High	Medium	Medium
$\mathbb{C}_C$	Low	Medium	High

The CoSC process takes in average 17 s per tile, and the computation time for the whole image ranged from 5 to 10 h depending on the parameters.

A badly chosen initial segmentation S leads to increased computation times and aberrant cases may result in bad segmentation and classification results. Finally, our first experiments suggest that the classification model $\mathbb{C}_C$ have no significant influence on computation time, but is crucial to segmentation and classification results. However this behaviour has to be verified by a more formal study which is out of the scope of this article. The impact of these parameters is briefly summarized in Table 2.

4.2 Results

In order to validate the applicability of CoSC, we compare our results to a pixel-based method, a classic OBIA approach and a hybrid segmentation-classification approach.

The pixel-based method consists in computing the normalized difference vegetation index (NDVI) (Rouse et al. 1974) which is a well-known radiometric index for the extraction of vegetation zones. The NDVI image is then normalized in [0, 1] so that we can see each pixel value as the probability of belonging to the vegetation class. In the rest of the article we denote this NDVI-based approach simply as NDVI. The classic OBIA method consists in classifying the initial segmentations with the same linear regression model that we used for CoSC. The hybrid segmentation-classification approach is a variant of the Spectral-Spatial Classification (SSC) approach described in Sect. 2. We evaluate the results in terms of classification and segmentation separately.

Classification

We computed both crisp and fuzzy quality metrics against the reference data (Fig. 2b). A threshold of 0.5 was used to *defuzzify* these data to compute the following crisp measures: accuracy (Acc), Precision (Pr), Recall (Re), F-measure (F1), and Cohen's kappa (κ). Table 3 shows the crisp results. We observe that the NDVI achieves great accuracy but the percentage of rejected or ambiguous pixels is 89%. The OBIA classic approach as well as the proposed CoSC reach acceptable Acc, Pr, Re, F1 and κ values. We remark that they reduced the percentage of ambiguous pixels to 52%

Table 3 Crisp quality measures

Image	Acc	Pr	Re	F1	κ	Am%	Q_{cs}
NDVI	0.997	0.999	0.988	0.994	0.992	0.894	0.103
OBIA	0.931	0.953	0.948	0.950	0.834	0.516	0.479
SSC	0.894	0.872	0.912	0.892	0.787	0.000	1.000
CoSC	0.930	0.952	0.943	0.947	0.844	0.464	0.531

Table 4 Fuzzy quality measures

Image	$\widetilde{Acc}$	$\widetilde{Pr}$	$\widetilde{Re}$	$\widetilde{F1}$
NDVI	0.774	0.791	0.720	0.754
OBIA	0.809	0.745	0.915	0.830
SSC	0.869	0.849	0.886	0.867
CoSC	0.825	0.768	0.912	0.834

and 46% respectively which is a good trade-off. SSC achieves high accuracy results without reporting ambiguous segments as the SVM model has binary output.

We also computed the fuzzy error matrix (Binaghi et al. 1999) and derived from it the fuzzy $\widetilde{Acc}$, $\widetilde{Pr}$, $\widetilde{Re}$, and $\widetilde{F1}$. Fuzzy metrics allows to take into account the intrinsic uncertainty of the reference data and have a more accurate quality assessment. Table 4 shows the fuzzy results for the compared methods. We observe that the NDVI method is globally outperformed by the other methods, and SSC achieves better classification results. There is no significant difference between the OBIA and CoSC methods, and they still have an acceptable performance.

Segmentation

Let's take a closer look at segmentation results. Figures 4, 5, 6 and 7 illustrate in detail the results of the NDVI method, OBIA, SSC, and CoSC respectively for the three tiles highlighted in Fig. 2a. The first row shows the probability images resulting from each method. The second row shows the boundaries of the segmentation corresponding to the flat regions (i.e., adjacent pixels having the same values) in the probability image.

We observe that the NDVI method is able to highlight vegetation zones quite accurately. However, the probability image is noisy so the flat-region based segmentation is extremely over-segmented; this shows one of the disadvantages of purely pixel-based approaches. Indeed, almost each pixel has a different NDVI value which results in the extreme over-segmentation observed in Fig. 4. More sophisticated post-processing techniques involving thresholds or clustering for example are required in order to obtain a good quality segmentation based on the NDVI approach.

The classic OBIA method achieves also a good discrimination of vegetation areas. However the associated segmentation is still over-segmented at many places. Indeed, in tile A, we observe that the different buildings are as over-segmented as in the initial

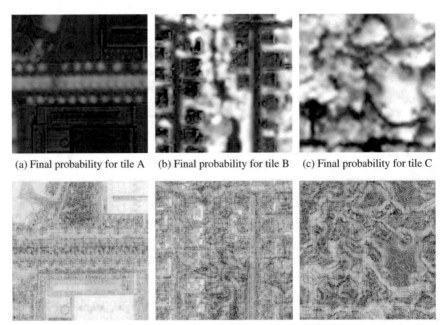

(a) Final probability for tile A   (b) Final probability for tile B   (c) Final probability for tile C

(d) Final segmentation for tile A (e) Final segmentation for tile B (f) Final segmentation for tile C

Fig. 4 Detailed view of the results obtained with the NDVI

(a) Final probability for tile A   (b) Final probability for tile B   (c) Final probability for tile C

(d) Final segmentation for tile A (e) Final segmentation for tile B (f) Final segmentation for tile C

Fig. 5 Detailed view of the results obtained with classic OBIA

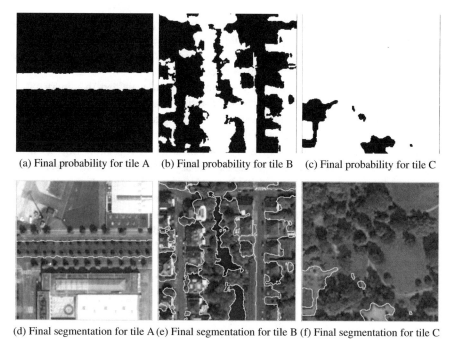

(a) Final probability for tile A   (b) Final probability for tile B   (c) Final probability for tile C

(d) Final segmentation for tile A (e) Final segmentation for tile B (f) Final segmentation for tile C

Fig. 6 Detailed view of the results obtained with spectral-spatial classification

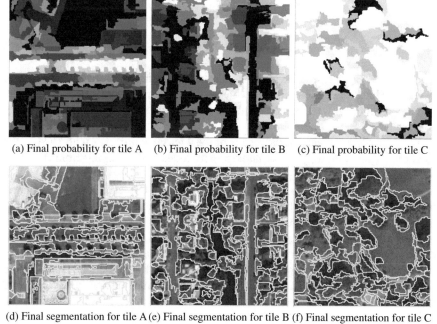

(a) Final probability for tile A   (b) Final probability for tile B   (c) Final probability for tile C

(d) Final segmentation for tile A (e) Final segmentation for tile B (f) Final segmentation for tile C

Fig. 7 Detailed view of the results obtained with CoSC

segmentation (Fig. 3d); in tile B we observe many elongated and irregular segments at the boundaries of some vegetation objects, also, the houses are still divided in many different segments; finally in tile C, we observe a lot of very small segments which could be merged to form medium sized trees.

The SSC method results in very under-segmented results. Indeed, the binary nature of this method as well as the post-processing filtering result in a very rough distinction between vegetation and non-vegetation objects losing all of the spatial information concerning higher level objects such as individual trees or buildings for instance.

By contrast, CoSC is able to give high probabilities to vegetation objects while giving very low probabilities to non-vegetation objects. Moreover, we observe that the associated segmentation is remarkably good. Indeed, there are almost no evident over- or under-segmented regions. Another surprising result is that the segmentation of non-vegetation objects was also improved as a side-effect. This can be explained by the fact that over-segmentation is mainly related to radiometrical homogeneity and the segmenter takes this into account to modify candidate segments, thus improving segmentation quality globally.

In order to objectively evaluate the segmentation results we employed the UOA_{L_2} metric. Recall that this metric is defined by:

$$\Psi = \sum_{R_i | \phi_\delta(R_i)=-1} \omega(R_i)$$

$$\Theta = \sum_{R_i | \phi_\delta(R_i)=1} \omega(R_i)$$

$$UOA_{L_2} = \sqrt{\Psi^2 + \Theta^2} \tag{6}$$

where Ψ and Θ represent the under- and over-segmentation rates respectively; and $\phi_\delta(R_i)$ is a local evaluation function which estimates if a segment is over-segmented, under-segmented or well segmented in function of a given homogeneity index H and a threshold δ.

For each tile we evaluated the results of the 4 tested approaches with UOA_{L_2}. We used the segment entropy as the homogeneity index and δ varying from 0 to 1 by 0.01 steps.

Figure 8 reports the results of these computations. Remark that UOA_{L_2} ranges from 0 (best) to 1 (worse); thus the lower the area under the UOA_{L_2} curve, the better the segmentation. For tile A and B, we observe that as expected the NDVI has the greatest area under the UOA_{L_2}, followed by SSC; CoSC and OBIA have very similar curves the area under the curve of OBIA is slightly lower than that of CoSC for tile A while for tile B we observe the inverse result. For tile C, the SSC method has the worst result, followed by the NDVI, the OBIA method and finally by CoSC which has a clearly lower area under the curve for this tile. These results confirm our previous observations, CoSC has indeed globally improved the segmentation, which is a very interesting result since the produced segmentation could be re-used for further processing and analysis in multi-class or multi-scale applications for example.

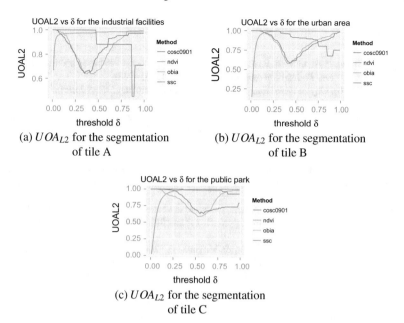

(a) UOA_{L2} for the segmentation of tile A

(b) UOA_{L2} for the segmentation of tile B

(c) UOA_{L2} for the segmentation of tile C

Fig. 8 Plots **a–c** show the variation of the UOA_{L_2} metric over threshold values in [0, 1] for the three tiles

5   Conclusion

In this article we presented CoSC, a collaborative framework for joint segmentation and classification. It aims at the extraction of a given thematic class of objects from a remote sensing image, and allows the exchange of implicit information between these different but closely related paradigms in order to simultaneously improve both segmentation and classification results. We presented the extraction of vegetation areas from a large image of Strasbourg as case-study to demonstrate the applicability and pertinence of the CoSC framework. Experiments show that the CoSC is able to give accurate classification results while remarkably improving the segmentation of the whole image, not only for the objects of interest but for all types of objects. In short-term we will study different strategies for choosing the candidate segment as well as different classification models and their influence on the results. The CoSC framework is designed for the extraction of a single thematic class; our future research will be focused on the study of collaborative strategies between many of such CoSC processes in order to achieve automatic and full (multi-class) interpretation of remote sensing images.

Acknowledgements The research leading to these results has received funding from the French *Agence Nationale de la Recherche* (Grant Agreement ANR-12-MONU-0001).

References

Binaghi, E., Brivio, P. A., Ghezzi, P., & Rampini, A. (1999). A fuzzy set-based accuracy assessment of soft classification. *Pattern Recognition Letters, 20*(9), 935–948.

Blaschke, T. (2010). Object based image analysis for remote sensing. *ISPRS Journal of Photogrammetry and Remote Sensing, 65*, 2–16.

Chen, H.-C., & Wang, S.-J. (2004). The use of visible color difference in the quantitative evaluation of color image segmentation. In *Proceedings of the ICASSP* (pp. 593–596).

Chow, C. (1970). On optimum recognition error and reject tradeoff. *IEEE Transactions on Information Theory, 16*(1), 41–46.

Derivaux, S., Forestier, G., Wemmert, C., & Lefèvre, S. (2010). Supervised image segmentation using watershed transform, fuzzy classification and evolutionary computation. *Pattern Recognition Letters, 31*, 2364–2374.

di Sciascio, C., Zanni-Merk, C., Wemmert, C., Marc-Zwecker, S., & de Beuvron, F. D. B. (2013). Towards a semi-automatic semantic approach for satellite image analysis. *Procedia Computer Science, 22*, 1388–1397.

Duveiller, G., Defourny, P., Desclée, B., & Mayaux, P. (2008). Deforestation in central Africa: Estimates at regional, national and landscape levels by advanced processing of systematically-distributed landsat extracts. *Remote Sensing of Environment, 112*(5), 1969–1981. Earth Observations for Terrestrial Biodiversity and Ecosystems Special Issue.

Farmer, M. E. (2009). *Application of the wrapper framework for robust image segmentation for object detection and recognition.* INTECH Open Access Publisher.

Hall, M., Frank, E., Holmes, G., Pfahringer, B., Reutemann, P., & Witten, I. H. (2009). The weka data mining software: An update. *ACM SIGKDD Explorations Newsletter, 11*(1), 10–18.

Haralick, R., Sternberg, S. R., & Zhuang, X. (1987). Image analysis using mathematical morphology. *IEEE Transactions on Pattern Analysis and Machine Intelligence (TPAMI), 9*(4), 532–550.

Hofmanna, P., Lettmayerb, P., Blaschkea, T., Belgiua, M., Wegenkittlb, S., Grafb, R., et al. (2014). Abia—a conceptional framework for agent based image analysis. *South-Eastern European Journal of Earth Observation and Geomatics, 3*(25), 125–129.

Kurtz, C., Passat, N., Gançarski, P., & Puissant, A. (2012). Extraction of complex patterns from multiresolution remote sensing images: A hierarchical top-down methodology. *Pattern Recognition, 45*, 685–706.

Lizarazo, I., & Elsner, P. (2011). Segmentation of remotely sensed imagery: Moving from sharp objects to fuzzy regions. *Image Segmentation.*

Mahmoudi, F. T., Samadzadegan, F., & Reinartz, P. (2013). Object oriented image analysis based on multi-agent recognition system. *Computers & Geosciences, 54*, 219–230.

Musci, M., Feitosa, R., & Costa, G. (2013). An object-based image analysis approach based on independent segmentations. In *Urban Remote Sensing Event (JURSE), 2013 Joint* (pp. 275–278).

Pham, H. M., Yamaguchi, Y., & Bui, T. Q. (2011). A case study on the relation between city planning and urban growth using remote sensing and spatial metrics. *Landscape Urban Plan, 100*, 223–230.

Promper, C., Puissant, A., Malet, J.-P., & Glade, T. (2014). Analysis of land cover changes in the past and the future as contribution to landslide risk scenarios. *Applied Geography, 53*, 11–19.

Räsänen, A., Rusanen, A., Kuitunen, M., & Lensu, A. (2013). What makes segmentation good? A case study in boreal forest habitat mapping. *International Journal of Remote Sensing, 34*, 8603–8627.

Rougier, S., & Puissant, A. (2014). Improvements of urban vegetation segmentation and classification using multi-temporal pleiades images. In *5th International Conference on Geographic Object-Based Image Analysis, Thessaloniki, Greece* (p. 6).

Rouse, J. W., Haas, R. H., Schell, J. A., Deering, D. W., & Harlan, J. C. (1974). *Monitoring the vernal advancement and retrogradation (greenwave effect) of natural vegetation.* Texas A & M University, Remote Sensing Center.

Tarabalka, Y., Benediktsson, J., & Chanussot, J. (2009). Spectral-spatial classification of hyper-spectral imagery based on partitional clustering techniques. *IEEE Transactions on Geoscience and Remote Sensing, 47*(8), 2973–2987.

Troya-Galvis, A., Gançarski, P., Passat, N., & Berti-Équille, L. (2015). Unsupervised quantification of under and over segmentation for object based remote sensing image analysis. *IEEE Journal of Selected Topics in Applied Earth Observations and Remote Sensing, 8*(5), 1936–1945.

Zhang, H., Fritts, J. E., & Goldman, S. A. (2003). An entropy-based objective evaluation method for image segmentation. *In Electronic Imaging 2004* (pp. 38–49).

Author Biographies

Dr. Andrés Troya-Galvis obtained the MSc at Université de Bourgogne, France in 2012. He obtained his PhD in computer science at Université de Strasbourg, France in 2016. He is now assistant lecturer at the engineering school INSA Strasbourg. His scientific interests include machine learning, image segmentation, data quality, and collaborative methods for remote sensing image analysis.

Prof. Pierre Gançarski is full Professor of Computer Science. He is affiliated to the Laboratory ICUBE (University of Strasbourg). His work focuses on collaborative multistrategy clustering, with application to remote sensing image analysis for urban and biodiversity applications. He proposes new approaches to a better exploitation of Earth Observation integrating multisources data to identify and monitor natural or anthropic environments. He has been the coordinator of different research projects on Collaborative Classification applied to Environmental Data.

Dr. Laure Berti-Équille is currently a Senior Scientist at Qatar Computing Research Institute (QCRI). Before, she was a Research Director ("Directeur de Recherche") at IRD, Institut de Recherche pour le Développement (2011–2013) and a tenured Associate Professor at the University of Rennes 1 in France (2000–2010). From 2007 to 2009, she was a visiting researcher at AT&T Labs Research (NJ, USA) with a Marie Curie fellowship. Her research interests range from data preprocessing, data cleaning, data integration, truth discovery to anomaly detection and exploratory data analysis.

Author Index

© Springer International Publishing AG 2018
B. Pinaud et al. (eds.), *Advances in Knowledge Discovery and Management*,
Studies in Computational Intelligence 732, https://doi.org/10.1007/978-3-319-65406-5

Printed in the United States
By Bookmasters